Angelo Heilprin

The Arctic Problem

And Narrative of the Peary Relief Expedition of the Academy of Natural ...

Angelo Heilprin

The Arctic Problem
And Narrative of the Peary Relief Expedition of the Academy of Natural ...

ISBN/EAN: 9783337008031

Printed in Europe, USA, Canada, Australia, Japan

Cover: Foto ©berggeist007 / pixelio.de

More available books at **www.hansebooks.com**

THE
ARCTIC PROBLEM

AND NARRATIVE OF THE

PEARY RELIEF EXPEDITION

OF THE

ACADEMY OF NATURAL SCIENCES

OF PHILADELPHIA.

BY

ANGELO HEILPRIN

LEADER OF THE PEARY RELIEF EXPEDITION

PROFESSOR IN THE ACADEMY OF NATURAL SCIENCES; PRESIDENT
OF THE GEOGRAPHICAL CLUB OF PHILADELPHIA.

PHILADELPHIA.
CONTEMPORARY PUBLISHING CO.
628 CHESTNUT STREET.
1893.

COPYRIGHT 1893,
Contemporary Publishing Co.

Engraved and Printed by
LEVYTYPE COMPANY,
Philadelphia.

Preface.

The interest which at the present moment centres about Polar exploration is perhaps broad enough to permit of a few additional pages being added to the lengthening literature of the subject, even though they be wanting in a recital of those mishaps and hardships which have made Arctic reading so fascinating. In this belief the author offers the following pages, which are in the main a record of personal experiences in the North, and reflections upon the best method of attaining the object which has so long baffled the energies of the hardy explorer. A portion of the work has already appeared in narrative form in the pages of Scribner's Magazine, and another portion is an amplification of an address delivered before the Geographical Club of Philadelphia.

The author feels that the record of the Peary Relief Expedition would not be complete without a reference to the numerous helping hands which made the expedition possible, and permitted of the full accomplishment of its mission;—to all these he owes a no small debt of gratitude, and to all, without distinction by name, he expresses his acknowledgments. A special mention should, however,

Preface.

be made of the names of a few gentlemen who more particularly interested themselves in the expedition, and gave their assistance in other directions besides the one very necessary one of raising the required funds for the undertaking. These are Gavin W. Hart, Esq., the indefatigable Treasurer of the Expedition Fund, and Messrs. Edward Longstreth, Joseph T. Rothrock, and Edwin J. Houston, through whose efforts, representing the good work of Philadelphia and West Chester, the rude children of the North have been placed in a condition of comparative comfort. The distribution of gifts of charity to the Eskimo was a feature of the expedition.

To the members of his party, for the faithful accomplishment of their duties, and the good will which ever prompted their work, the leader is placed under special obligation; and he is similarly indebted to the officers and crew of the good ship Kite, the vessel of the expedition.

A. H.

Philadelphia, May, 1893.

CONTENTS.

I. The Arctic Problem, ... Page 5.
I. Polar Expeditions, ... " 28.
III. The Spitzbergen Route to
  the Pole, ... " 53.
IV. The Peary Relief Expedition, " 79.
V. A Lost Companion, ... " 128.
VI. The Greenland Ice-Cap and
  its Glaciers . . . . " 114.

"SAIL on, nor fear to breast the sea!
   Our hearts, our hopes, are all with thee,
Our hearts, our hopes, our prayers, our tears,
Our faith triumphant o'er our fears,
Are all with thee,—are all with thee!"

I.

THE ARCTIC PROBLEM.

It can truly be said that to no class of workers does a community owe so much of what it possesses as to that of the travelling geographer or explorer, whose cumulative efforts and resources have brought the world to know itself as it in reality exists. This year the nations of the world celebrate the greatest discovery of modern times,—the greatest in the sense that it most deeply touches the welfare of the human-kind. Every year, almost, brings with it the culmination of an effort which, though not so great and far-reaching as the Columbian, yet adds materially to that fund of knowledge from which Columbus drew his inspiration, and which served as the mainspring for the discovery of a continent. On sea and on land, from alpine summits to the waters of the frozen north, the march of discovery is progressive, and it will forever remain progressive and unabated until the surface of our globe is made known to us in its every feature and under every phase of its existence.

The region which to-day again most attracts the thoughts of geographers lies in the far north. Like the pulse which alternately stills and throbs with each changing phase of physical depression and elevation, the Arctic pendulum, held and swayed by every record of disaster or success, vibrates between periods of peaceful slumber and restless activity. Almost every decade of the present century has seen the ice-field of the north attract to it the energies of men who have worked in the cause of humanity and have made exploration brilliant; every equal period, almost, has brought with it its chapter of disaster and failure. Not daunted by the opinion of those who hold that man's legitimate field of labor lies only there where it can return immediate profit to one's ownself or to the community at large, the wayfarer to the Arctic seas plods wearily onward, slowly but steadily closing those gaps of knowledge of which he has been appointed the trusted guardian. Without him the world would know less and be a corresponding sufferer in its lack of knowledge.

The brilliant success which has attended Mr. Peary's recent traverse of the Greenland ice-cap opens up anew the question of the accessibility of the Pole, and the discussion of this subject is specially apropos at this time on the eve of the departure of two expeditions whose destination is equally the North Pole, and of a third, from which, while not in quest of the Ultima Thule of most Arctic ventures, it can reasonably be expected that a higher latitude will be reached than has ever before been attained. The first expedition, under command of Fridtjof Nansen, the preparations for which have now nearly been completed, assumes for itself the route of the north Siberian waters and the help of a slow and steady drift of the ice-pack northward from the Eurasiatic continent to and across the Pole, with a return to the Spitzbergen or Greenland coast. In other words, it is the purpose of a vessel, specially constructed to resist the pressure of the ice, to enter the ice-pack, not far from the spot where the unfortunate Jeanette was crushed in 1881, and allow itself to be drifted by the pack, in its own course, and free from the disturbing influence of a navigator, for a period of some two or three years. The assumption that there is a steady currental drift northward from the Siberian waters is based almost entirely upon the records which a few pieces of clothing and other relicts have carried with them of the ill-starred expedition of 1879, and the circumstance that drift-wood, supposed to come from Alaska, is annually thrown upon the east Greenland coast. The Jeanette relicts, to which reference has just been made, consisting of the trousers of one of the sailors, and of parts of the ship's papers, were picked up on the ice of Julianehaab, southwest Greenland, on June 18th, 1884, eleven hundred days after that unfortunate disaster which marks one of the saddest chapters in the history of Arctic exploration. The drift, whatever its course may have been, consumed three full years; to meet the possibilities of the same period and of a needed prolongation of time, Nansen has provisioned himself for a period of five years. It is impossble to foretell the fate of this expedition; its success or failure will depend upon a combination of circumstances and conditions which cannot even be premised in our present knowledge of the northern seas. The known perseverance and indomitable pluck of the leader lend hope for the enterprise, if not absolute assurance of success.

The Ekroll Expedition, which is expected to leave Cape Mohn, on the east coast of Spitzbergen, early in June, 1893, is a revival in one sense of the expedition of Parry, in 1827, when the very high latitude of 82° 45′ was attained. Its special feature is the construction of a "combination" conveyance to be used alternately in boat and sledge service—*i. e.*, a boat that at the needed time might be resolved into a number of sledges. The virtual point of departure of this expedition is Petermann's Land, the off-lying island situated north of Francis-Joseph Land, whence a direct traverse is contemplated to the Pole, with a return, if possible, by way of Fort Conger, in Lady

Franklin Bay. Supan has constructed the following table of distances for the expedition:

From Cape Mohn to Petermann's Land, 700 km.
" Petermann's Land to the Pole, . 830 "
" the Pole to Fort Conger, . . . 950 "
_____
2480 km.

or approximately 1500 miles. Two hundred days are allowed for the accomplishment of this journey.

With many, probably, the Ekroll Expedition, ingeniously though it may be planned, will not fully lend itself to favor, especially in its dominant feature, the construction of a sectioned boat for alternate boat and sledge service. The interdependence of the two constructions measured as a factor of safety is an element of insecurity, or at least uncertainty, in the enterprise which should be carefully weighed and considered before the initial stages in an undertaking so hazardous as the one here contemplated are taken. One may perhaps, express a further doubt in the matter of expediency; the constant shifting of the impedimenta of travel is a condition which should be avoided so far as is possible in a journey, the execution of which is dependent almost wholly upon a reserve strength and that adjustment of labor and rest which permits of the greatest amount of bodily and mental vigor being maintained. The main work of any Polar expedition should be to go ahead, and not the labor of adjustment and rearrangement looking only toward the accomplishment of this end.

* Mr. Peary's new expedition contemplates the further exploration of the north and east Greenland coasts, to the extent of surveying the still unknown, or at best conjectural, boundary which unites Cape Bismarck on the east with the extreme north and with the "farthest" of Lockwood and Brainard on the northwest (in latitude 83° 24'). This course involves the circumnavigation, or more properly—since the effort will doubtless be executed on the frozen surface of the sea—rounding of the Greenland archipelago, the insular masses lying north of Independence Bay and of what appears to be a continuation of the Victoria Inlet. To what degree of northern latitude this course of exploration may take, cannot in the nature of things be predicted; but there are reasons for believing that the outlying insular masses extend to fully the 85th parallel, and possibly much further. Carried out on the lines of the exploration of the past year, with the inland ice-cap as the main line of travel, there are good grounds for hoping for an amount of success equal to that which has made the year 1892 memorable in Arctic service.

No doubt the eyes of the world will follow with interest the experiences of these daring

navigators of the north. The question will, however, certainly be asked: To what good? For those who identify the progress of civilization with a search after truth, it is hardly necessary to answer this question. It was fully answered a half-century ago by that stern friend of knowledge, Sir John Barrow, when he wrote; "The North Pole is the only thing in the world about which we know nothing; and that want of all knowledge ought to operate as a spur to adopt the means of wiping away that stain of ignorance from this enlightened age." Alexander, after he had conquered the world, is said to have grieved because he had failed to accomplish the ambition of his life,—the discovery of the sources of the Nile—a legacy which was bequeathed to his successors through a period of 2200 years. Whatever protest might be indulged in to prevent further Arctic exploration, whatever specious reasons advanced for not "further risking the lives of more able men," it can be accepted as a certainty that until the region of the Pole is in fact traversed, or its inaccessibility absolutely demonstrated, the pendulum of Arctic exploration will continue to swing. The search after knowledge has no limits and it knows no time.

It is customary to decry Arctic exploration on the sole ground that it endangers the lives of worthy people, while it yields little of benefit in return. This is the narrow view of that vast body of opinionists who from habit prefer to say much to thinking little, who know the world from their own knowledge, rather than from the knowledge obtained by others. The successful man is in their eyes a hero, the unfortunate one, although travelling in the same field of honest labor or research, scarcely better than an imbecile. The exploits of a Ross, a Kane or a Hayes are held up for emulation—those of a Franklin, Hall, or DeLong for condemnation. All of these persons strove for a common purpose, and from each the world has derived an almost equal share of profit. Each of these heroes and others before and after them have contributed to that special store of knowledge which conquers assumed impossibilities—which renders practicable much that had before been considered impracticable.

It is not difficult to discern the practical results or benefits arising from Arctic exploration. The location of the Magnetic Pole alone, rendering possible the determination of the lines of variation in the magnetic needle, is in itself a conquest for which navigators will for all time be grateful, and from which the world at large has derived inestimable benefits. The Arctic whale fishery is principally an outcome of Arctic exploration, made practicable and profitable through that more intimate knowledge of the physical conditions of the far north which has been begotten alike of the labors of success and disaster. Every expedition, almost, has accomplished something that had been left undone by its predecessor and been considered in the nature of things unattainable. The brilliant exploits of Lockwood and Brainard, when they penetrated to within five hundred miles of the Pole, or to Lat. 83° 24′; of Nordenskjöld, the discoverer of the North-East Passage; and of Nansen, the transgressor of Greenland's supposed impassible ice-fastnesses, are a chapter in the history of the present decade. Among its pages, too, will be written the narrative of that remarkable journey which has only recently been accomplished and which in boldness of execution outranks the achievements of all previous explorers in the same field.

It may, however, be pointed out that the exploits to which reference has just been made are of insignificant import; that Parry as early as 1827 had reached a point north within forty-five miles of that attained by Lockwood and Brainard; that McClure, despite his brilliant forcing of the North-West Passage, had yet failed to render commercially navigable the route in the search for which Sir John Franklin and the greater part of his force gave up their lives. Where is the profit? The contention is just, or better true, but only in so far as the simple statement of fact is concerned. Parry's farthest north bears certainly the mark of a brilliant achievement, but yet the position is not so far north of the point where Greely and his party spent the better part of three years as to disguise the real advance that has been made in Arctic exploration. Hall, Markham, Aldrich, and Lockwood and Brainard have all passed beyond Parry's "farthest", and it is a well-known fact that almost every expedition advances upon its predecessor. Again, with reference to the "barrenness" of the North-West Passage: What do we know of its possibilities—only the record of failures? Mainly so; but is the experience of a few Arctic ventures, most of them badly conducted or ill-arranged, to be taken as the guiding line on which the possibilities of the future are to be weighed? Our mental equipment is such that we almost invariably judge of possibilities in the light of existing knowledge; in geographical exploration, as in all departments of mechanical and physical science, however, it has repeatedly been shown that the assumed impossibilities of one day are ready possibilities of another, and that there are no fixed limits in which the element of success can be determined. The heroic achievement of Paccard, who in 1786 first scaled the then seemingly inaccessible summit of the Mont Blanc, is to-day scarcely remembered, so facile—one might almost say, fashionable—has become the route along which the first breach was effected. Humboldt's ascent of Chimborazo added lustre to the researches of that remarkable investigator, but to-day, after what has been accomplished by the brothers Schlagintweit, by Graham and Conway in the Himalayas, by Donkin and Freshfield in the Caucasus, by Meyer on Kilima 'Njaro, by Güssfeldt on Aconcagua, by Reiss, Wolff, and Whymper among the Equatorial Andes, and by Russell on St. Elias, such an undertaking would scarcely pass beyond the records of the geographer and the archives of geographical societies. Neither the sands nor the swamps, or even the dark and gloomy forests of the deep interior, can any longer reasonably be counted upon to thwart the purposes of the African traveler. Stanley's remarkable traverse of that continent on what might be termed schedule time, and Emin Bey's equally remarkable sojourn of years in a region approach to which had for a long time baffled the energies of men of most undoubted courage, are an evidence of direct evolution of possibility from knowledge and experience. Similarly, in the far north, the dreaded dangers of Melville Bay can to-day, with proper judgment, be avoided with much the certainty that the dangers of the fog-banks are avoided by the regular trans-Atlantic liners.

We are as yet too ignorant of what the north

promises to permit us to venture upon a statement of the possibilities which it offers to either commerce or science, but certain it is that its inaccessibility is becoming more and more remote every year. Albeit the North-West Passage has not yet proved of commercial significance, who can predict what its future might not be? Equally unpromising has seemed the passage, only once effected, in the opposite direction, but the explorations of Nordenskjöld are already beginning to bear fruit. The successful issue of this journey has revived the so-called "north Siberian trading" route, and the day appears not far distant when it will be freely used as the direct means of commercial communication between the north of Europe and north-central Asia. The successful ventures of Captain J. Wiggins in 1888 and 1889, when with little delay he reached the mouth of the Yenissei River, and of Peterson, Cordiner, and R. Wiggins in 1890, seem to justify the hopes that have been held out for the new route, and to bring promise, at least, for the "Anglo-Siberian Trading Syndicate."

There are two phases to the exploration of the region of the far north: that which touches exploration only in the broader sense of a contribution to scientific knowledge, and the other which has for its first and ultimate object the attainment of the position of the North Pole. It has been contended with a certain amount of force that from a scientific or geographical standpoint the Pole offers nothing more of interest than would be offered by any other unknown point of the far north, and that, consequently, exploration toward it should treat the Polar question merely as an incident rather than as an object. This is not strictly the case, however; for from the simple fact that the Pole has thus far proved inaccessible, the question of "why this inaccessibility"— the determination of the conditions which make this approach seemingly impossible— must forever be of special scientific moment, and until the question is definitely answered by a practical demonstration, or once and forever removed from the field of possible answer, it will continue to attract to it the minds of the speculative as well as of those who are concerned only with an immediate result. Arctic exploration, in the Weyprechtian sense of an exploration for the attainment of scientific knowledge pure and simple, is worthy of all the effort that can be put to it; but none the less worthy is an exploration which has for its main object the resolution of a problem which has attracted man's attention for upwards of three hundred years, and thus far baffled all his ingenuity and advances. It might almost be said with Sir Martin Frobisher, who wrote toward the close of the sixteenth century (or nearly three centuries in advance of the actual discovery) regarding the North-West passage: "It is the only thing in the world that is left yet undone whereby a notable mind might be made famous and fortunate."

Two questions here naturally suggest themselves: Is the Pole in fact accessible, and if so, how and by what route is it to be attained? As regards the first question, the answer can be freely hazarded that with the advances in the art of travel that have latterly been made —and it is needless to conceal the fact that to the explorations of the past year we owe more in this regard than to any preceding exploration—its conquest will be assured before many years have passed. As to the route to be followed, conjecture has broad scope and an ample field.

With two exceptions—the Austro-Hungarian Expedition of 1872-74 and the Herald-Jeanette Expedition of 1879—all the more notable efforts made within recent years to attain the "farthest north" have been by way of the west Greenland waters *i. e.*, Baffin Bay, Smith Sound, and Kennedy and Robeson Channels. The Austro-Hungarian Expedition was imprisoned in the ice of Francis-Joseph Land, while the Jeanette was crushed in the ice of the north Siberian waters. The two Polar expeditions that are maturing their plans for work in the present year will follow largely in the paths of the Tegethoff and Jeanette.

The estimation of the possibilities of reaching the Pole can only be calculated on the basis of past failures. What improvements or changes in the method of travelling suggest themselves so as to convert failure into success? It now seems to be conclusively proved that any attempt to sail to the Pole by way of the west Greenland seas must be doomed to failure; the open Polar Sea, which has so largely figured in Arctic ventures, has vanished—at least from the American hemisphere—and in its place is a seemingly interminable barrier of ice, the ice of the so-called Palæocrystic Sea. An effort to penerate this frozen sea in the direction of the Pole was made by Markham in 1876, with the result of attaining the high latitude of 83° 20'. The able commander of the sledge-journey states: "I feel it impossible for any pen to depict with accuracy, and yet be not accused of exaggeration, the numerous drawbacks that impeded our progress. One point, however, in my opinion is most definitely settled, and that is, the utter impracticability of reaching the North Pole over the floe in this locality; and in this opinion my able colleague, Lieutenant Parr, entirely concurs. I am convinced that with the very lightest equipped sledges, carrying no boats, and with all the resources of the ship concentrated in the one direction, and also supposing that perfect health might be maintained, the latitude attained by the party I had the honor and pleasure of commanding would not be exceeded by many miles, certainly not by a degree."

While one cannot but admire the courage of this statement, emanating as it does from one of the most tried of Arctic explorers, it can nevertheless scarcely be considered to have proved its case. There is as yet nothing to prove that the condition of impassability of the Palæocrystic ice is the same year for year; the observations of one season are not sufficient to demonstrate this, as the experiences of Lockwood and Brainard in the spring of 1883 clearly prove. These intrepid explorers in their second effort to reach the farthest north found the Polar pack disrupted and moving already in the early days of April, and at a point not much beyond the 82d parallel. The conditions of the previous year had been entirely reversed. And if the conditions of the ice change, do not also the possibilities of travel? Again, the extraordinarily rapid sledge journey of Lockwood and Brainard in 1882, when they attained the most northerly point ever reached by man, 83° 24', clearly points out a line of possibility which was certainly not appreciated by the officers of the British expedition. Who can tell how far Lockwood and Brainard might not have penetrated had they been furnished with supplemental time and that form of experience, including its resources, of which Mr.

IN THE WAIGAT.

Peary took such good advantage? There is reason to believe that they might have travelled 200 miles or more further, and possibly even reached the Pole in an early and favorable season. From Lady Franklin Bay, the headquarters of the Greely Expedition, the distance north to the Pole is less than six hundred miles; at the average rate of progress made by Lockwood and Brainard for a period of twenty-two days, from Cape Bryant to Lockwood Island and return, a direct traverse to the point of the earth's axis would be made in forty days; or allowing twenty-five per cent for deflection from a due north and south traverse, in fifty days. Payer, in his second sledge journey in April 1874, accomplished the return of 313 miles (latitudinal distance) in thirty days, under the most exhaustive conditions of hummocky and mountainous ice, drift snow, an unnecessarily heavily laden sledge, and with a temperature descending to $-22°$ F. The late Greenland expedition traversed 1300 odometric miles in ninty-six days, or an average, including all stoppages, of thirteen and a half miles per day. But in reality the greater part of the return journey was made at the astonishingly rapid rate of 20—25 miles, and some of it at thirty miles a day.

It is true that the special conditions which favor travelling on the Greenland snows are not to be met with on the rough surface of the frozen sea, but even with this difference, and recognizing as permanent the conditions which Markham emphasizes, there seems to be no good reason why, with a properly adjusted equipment, the Pole might not be reached. In all the sledge expeditions thus far, with the exception of Mr. Peary's, progression has been very largely effected on the principle of "doubling up"—*i. e.*, drawing the load in sections, and thus retraversing the same ground two, three or even five times over. The unnecessarily heavy equipment necessitated this method of advance. Lockwood's sledge, drawn by eight dogs, and equipped with the impedimenta for three men for twenty-five days, weighed 783½ lbs.; Payer's sledge, equipped for seven men and three dogs for twenty-eight days, weighed 1565 lbs.; Parry's memorable sledge journey north of Spitzbergen, in 1827, was made with two equipments of 3800 lbs. each, the provisioning being for seventy-one days serving twenty-eight men; the load calculated for each man was 270 lbs. Mr. Peary's equipment, on the other hand, on leaving the Humboldt Glacier, or seventy-seven days before the close of the journey, weighed only 1500 lbs., which was distributed exclusively among the dogs pressed in for service. With this comparatively light load the travel was made direct and without doubling.

The traverse to the Pole—barring a possible

direct passage through the "pack sea" by a powerful steamer—owing to ice-driftages and water holes, can seemingly be made only by a combination of boat and sledge journey, and that it can so be accomplished with the use of a light equipment, admits of little or no doubt. Parry's remarkable journey of 1827 is almost conclusive evidence on this point. This most cautious and circumspect navigator, in his memorable journey northward of twenty-nine days, accomplished 292 geographical miles, largely over soft and floating ice; or, with the journey repeated, as it had to be, three and not unfrequently five times over, the actual space covered was, as stated by Parry himself, approximately 668 statute miles, "being nearly sufficient to have reached the Pole in a direct line." It is one of the anomalies of Arctic exploration that the route which Parry broke into has never since been followed. The thoroughness with which his expedition had been conducted and its failure to reach the point desired, seem to have been sufficient to condemn the route for all time. However wise or unwise the abandonment of the Parry route may have originally been, it is all but certain that with the Arctic experience that has been acquired during the last twenty-five years, and with such special facilities for travelling as have become incorporated in the "art" during the same period, the route in question

could be advantageously followed to-day. Indeed, Parry himself, commenting upon his own failure eighteen years after the event, gives it as his opinion that the object is "of no very difficult attainment, if set about in a different manner." He believed that starting from the north of Spitzbergen in the month of April, "when the ice would present one hard and unbroken surface" "it would not be difficult to make good thirty miles per day without any exposure to wet, and probably without snow blindness."

The advantages which a condensed food supply, proper body-covering, and light constructions carry with them—advantages unknown to the early Arctic explorers—are a factor in the calculation which tends wholly to the side of success.

If it be asked why Parry failed to reach a higher point than he did, the answer is immediately found in his narrative: the southerly drift of the ice annihilated the actual northing made. But it is manifest that with the ice in a condition of stability, as it probably largely is at an earlier season, and with the equipment so adjusted as to obviate the necessity of "doubling" over the line of traverse, a much greater daily advance could be made than Parry found possible, and one that would three or four times cover the four-mile drift of the ice. With a daily advantage of only eight miles the Pole would be reached from the 82d parallel in sixty days; while on the other hand, the return journey with the drift would probably not occupy much more than half that period of time.

A further question suggests itself in connection with the Arctic problem: Has the Pole ever been more nearly accessible than it is at the present time? Without entering into a discussion of the geological problem or to a consideration of remote periods of time, it can be said that we are in possession of a certain amount of evidence which goes far toward giving an affirmative to the question. We owe chiefly to the Honorable Daines Barrington, that indefatigable advocate of Polar expeditions of the last century, the collection of data bearing upon this point. From these it would appear that a century or more ago voyages to the far north were not an exceptional circumstance—indeed, that the latitudes attained on these voyages were in some instances beyond what it has been possible to attain since. Thus, it is asserted that Capt. MacCallam, in command of the Campbeltown, one of the ships employed in the Greenland Fishery, in 1751 penetrated to 83° 30' in a perfectly open sea, with no ice visible to the northward. In 1754 three whalers are reported to have passed beyond the 82d parallel (to 82° 15', 83° and 84° 30'); the circumstances connected with the voyage of Mr. Stephens, when the latitude of 84° 30′ is said to have been attained, are reported by the Astronomer Royal of the time, Mr. Maskelyne, and are so circumstantially stated that they would seem to allow little room for doubt in the manner. But little ice was seen or met with beyond Hakluyt Headland. The same condition was reported in 1766 by Capt. Robinson, who claims in that year to have reached 82° 30′; and in the same year, along the same course, Master Wheatley reported no ice in any direction visible from the mast-head of his vessel in Lat. 81° 30′.

Unfortunately, the evidence that is presented is not of that kind which permits it to be accepted without reservation, but the coincidence as to years of easy passage—1754, 1766, 1770—and of a generally facile period about that time is certainly suggestive of an amount of truth-color. It is evidence of a kind that cannot be brushed aside merely because it conflicts with notions formed on more modern experiences, or because it emanates from the non-erudite whaler. The popular notion of a change in the winter climate of the eastern United States, with a suggested explanation that it might be due to a transference nearer to the coast of some of the heated waters of the Gulf Stream—a notion ridiculed in many quarters of the scientific world—seems to have received a nearly full vindication through the recent investigations of Prof. Libbey. That in 1817, and again in 1818, the greater part of the ice-mass which as a rule, year in and year out, almost impassably bounds the northeast coast of Greenland, suddenly disappeared, is proved by the unimpeachable testimony of the younger Scoresby; at that time a largely unencumbered sea seems to have extended along the coast very nearly to, if not beyond, the 80th parallel. This circumstance makes it probable that more truth is contained in the statements of the old whalers than has been commonly allowed. Indeed, it would appear from Parry's own narrative that at the time of his expedition a staunch steam-vessel, such as is now used for purposes of this kind, could have forced a passage much to the northward of the farthest point reached by him, or far beyond the most northern point which has ever been reached by a vessel, whether steamer or sailer.

II.

Polar Expeditions.

The first attempt, of which we have record, to reach the Pole seems to have been made in 1527. In that year, at any rate, according to the Chronicles of Hall and Grafton (as quoted by Hakluyt), King Henry VIII, actuated by "very weighty and substantial reasons to set forth a discoverie even to the North Pole," "sent two faire ships well manned and victualled, having in them divers cunning men to seek strange regions, and so they set forth out of the Thames the 20th day of May, in the 19th yeere of his raigne, which was the year of our Lord 1527." Little is known of this expedition beyond the fact that one of the vessels, whether the Dominus Vobiscum or not, was lost in the waters north of Newfoundland, and that the other returned home about the beginning of October. A Canon of St. Paul's, in London, reputed to be a great mathematician and a wealthy man, was one of the party on the Dominus Vobiscum.

In 1596, in an attempt to make the trans-Polar passage from Amsterdam to China, William Barentz, Jacob Van Heemskerke, and Cornelis Ryp reached Lat. 80° 11′ (on June

19th), off what is with little doubt Amsterdam Island, Spitzbergen. It is not known whether further progress northward was absolutely prevented by an impassable barrier of ice or not. Failing to make the passage on the route selected Barentz, with one of the vessels, sailed eastward for the Waigatsch, hoping to be more successful on a lower parallel. In this hope, however, he was disappointed; his vessel was hemmed in by the Nova Zembla ice, and there abandoned to its fate. The fortunes of the expedition were then thrown into the two open boats. Passing a winter of unusual hardship, the party on the 13th of June following (1598), began that memorable retreat homeward which may be considered to be pioneer among similar Arctic mishaps—a monument to the "worthy stuff" out of which a ship's crew was made three hundred years ago. Barentz and one of his associates, Claes Adrianson, died a few days after the departure, but the remainder of the party successfully pulled through to Kola, where they were rescued by Cornelis Ryp. Barrow well says (1818): "There are numerous instances on record of extraordinary voyages being performed in rough and tempestuous seas in open boats, with the most scanty supply of provisions and water, but there is probably not one instance, that can be compared to that in question, where fifteen persons, in two open boats, had to pass over a frozen ocean more than eleven hundred miles, 'in the ice, over the ice, and through the sea,' exposed to all the dangers of being at one time overwhelmed by the waves, at another of being crushed to atoms by the whirling of large masses of ice, and to the constant attack of ferocious bears, enduring for upwards of forty days severe cold, fatigue, famine, and disease." The most noteworthy outcome of Barentz's expedition was the discovery af Spitzbergen.

The distinguished navigator Henry Hudson, in an attempt to make the trans-Polar passage in 1607, reached the very high latitude of 80° 23'. It is not known at precisely what point this position was attained and it is difficult to correlate the observations made with our present geographical knowledge of the region (the Greenland-Spitzbergen Sea). Hudson himself states that land was visible southward, and that it extended in the opposite direction "farre into 82 degrees." Barrow remarks in this connection that this statement must be erroneous, or that Hudson stood over so far to the westward as to again bring him into proximity with the Greenland coast ("Chronological History of Voyages into the Arctic Regions," p. 182). This conclusion was reached from the supposition that Greenland was the only land-mass in the region under consideration which extended to or beyond the

A QUIET DAY ON MELVILLE BAY.—THE "ICE BLINK."

81st parallel; the since-discovered Francis-Joseph Land, however, makes it possible that Hudson's course was northeastward, rather than northwestward, and that the land reported to have been seen was in fact the disjointed tract which was reached by Payer and Weyprecht in 1873. Two considerations seem to sustain this view: first, the more ready accessibility of the coast of Francis-Joseph Land, so far as an unencumbered sea is concerned, and secondly, its own comparative proximity. One day's sail from the northern extremity of Spitzbergen would almost bring the mountain masses of Francis-Joseph Land into prominence, "trending," as stated by Hudson, "north in our sight." On the other hand, a voyage of four hundred miles westward would be necessitated before any part of the Greenland coast lying north of the 80th parallel could be brought into view; again, this portion of the Greenland-Spitzbergen Sea has generally been found to be impassable by reason of the heavy accumulation of drift ice. No landing on the Greenland east coast has thus far been found possible north of Cape Bismarck (77° 1'), the position reached by Koldewey in 1870, yet Hudson narrates that (in 80° 23') his men were successful in making a landing, that "they found it hot on shoore, and drunke water to coole their thirst, which they also commended." It is possible that the summer season of 1607 was an open one, just as Scoresby found it in 1817, and that the Greenland ice had disappeared even from the region of the 80th parallel; in that case Hudson may have approached to, or even landed on, the Greenland coast, and indeed, a suspicion of this condition is found in a (subsequent) passage which refers to "neerenesse to Groneland," and in his desire to "returne by the north of Groneland to Davis his Streights, and so for England."

If, as is claimed, land was visible far to the northward of the 82d parallel, then, manifestly, Hudson must have penetrated considerably beyond 80° 23', as it is wholly unlikely that he saw any land at a greater distance than 80 or 100 statute miles; at any rate, this could not have been on the Greenland side, where, as we now know, the coast beyond the 80th parallel trends rapidly westward. A mountain 4,000 feet in elevation, under the most favorable conditions, can be seen from the sea-level from an extreme distance of only eighty miles; and from a masthead one hundred feet in height, about twelve miles further.

Another effort in the direction of the Pole was made by a certain Jonas Poole, master of the ship Amitie, in 1610, but the farthest latitude reached (off Spitzbergen) was 79° 50'. The commission of this voyage, undertaken under the auspices of the Muscovy Company, recites: "Inasmuch as it hath pleased Almightie God, through the industry of your-selfe and others, to discover unto our nation a land lying in eightie degrees toward the North Pole: We are desirous, not only to discover farther to the northward, along the said land, to find whether the same be an island or a mayne, and which way the same doth trend, either to the eastward or to the westward of the pole; as also whether the same be inhabited by any people, or whether there be an open sea farther northward than hath been already discovered. . . . ."

The same Jonas Poole seems to have attained to about the 80th parallel in the following year, while Robert Fotherby, three years later, reached the northeastern extremity of Spitzbergen, in Lat. 80° 16'. Between this period and 1773, when Captain Constantine John Phipps (afterwards Lord Mulgrave) was entrusted with the command of an expedition to determine the limits of navigation northward, there seems to have been no "official" or organized effort to attain any very high latitude. It is true that in the instructions given to Baffin and Bylot to guide the expedition of 1616 in search of a North-West Passage we find the following: "For your course you must make all possible haste to the Cape Desolation; and from thence you, William Baffin as pilot, keep along the coast of Greenland and up Fretum Davis, until you come toward the height of eighty degrees, if the land will give you leave;" but this northern diversion seems to have been in no way associated with the thought of an actual northerly passage, since the instructions further recite that the voyage is thence to be conducted westward and southward, with a fall to the 60th parallel. The exact position of Baffin's "farthest north" is not known, but it was doubtless beyond the 78th parallel, probably just within the Kane Basin. Phipps reached positions due north of Spitzbergen of 80° 34', 80° 37', and 80° 48' (July 27th), at which points the great ice-barrier trending in an almost due east-and-west line prevented further penetration northward.

The failure to reach a more northerly point is significant, since it is alleged that in the same year Captain Clarke sailed to 81° 30' and Captain Bateson to 82° 15'. Scoresby, who seems inclined to distrust the accounts of these whalers, thus forcibly expresses himself in this connection: "Now this was the year in which Captain Phipps proceeded on discovery, towards the North Pole, who, notwithstanding he made apparently every exertion, and exposed his ships in no common degree; though he repeatedly traced the face of the northern ice from the longitude of 2° E., where the ice began to trend to the southward, to 20° E., where he was so dangerously involved, was never able to proceed beyond 80° 48′ N., and even that length once in the season. Is it reasonable, therefore, to suppose that whale-fishers, sailing in clear water, without any particular object to induce them to proceed far towards the north, should exceed the length to which Captain Phipps attained in the same year, and within a few days of the same time, by eighty-seven miles towards the north? I imagine, on the contrary, that both Captain Clarke and Captain Bateson had been mistaken in their latitude, and had not been so far as Captain Phipps, or at least not farther."

Scoresby's objections are well taken, yet the rapidity with which the northern ice frequently breaks, and as rapidly closes over, as I have myself had occasion to observe during two seasons, makes it not exactly impossible that with the difference in time of a few days the whalers may have been actually successful where Phipps was not; again, the recklessness —a characteristic which, at least to-day, marks that class of people—of the whalers might have impelled to a course which would not have been considered expedient to an officer of trained responsibility, such as Phipps. It is well known that in the west Greenland waters, in Melville Bay, the whalers, in their anxiety to be first on the American whaling "ground," regularly subject their vessels to the chances of crushing by forcing ("butting") a passage through the ice, rather than wait for it to disappear in its normal course. In this way the traverse is frequently made in the early days of June—sometimes still earlier—when the sea is seemingly still one impassable ice-pan, whereas the so-called "open passage" is usually not made before the middle or last week of July. In our traverse of the Bay in 1891, under the guidance of an experienced, but perhaps too cautious, ice-master, we were detained in the ice until the 23rd of July—becoming imprisoned on the 2d; yet in the following year, Captain Phillips, of the whaler "Esquimaux," who kindly offered to apprise Mr. Peary in advance of the coming of a relief expedition, had already as early as the 13th of June succeeded in planting my message on Wolstenholme Island, fully fifty miles beyond the gates of Cape York.

It is not then unreasonable to assume that the hardy and venturesome whaler will frequently penetrate where the dictates of caution would restrain the disciplinarian, or man of more impressed responsibility; therefore, due weight must be given to this consideration in judging of the possible accomplishments of the two classes of navigators. That a much further northing than Phipps's had already been made some time before the voyage of that navigator is almost indisputably proved by the Dutch charts of 1707, which place a "Gilles' Land" on the 82d parallel, and in a position which, in its relations to the trend of Zichy Land (Francis-Joseph Land), make it all but certain that land had in fact been discovered at that time.

In order to encourage exploration in the far north the English Parliament in 1776, by special enactment (16th Geo. III., chap. 6), offered a reward of £5,000 to the owner of any merchant vessel or to the commander of any King's vessel who should first penetrate to within one degree of the Pole, a fitting supplement to the act of 1743 (16th Geo. II., chap. 17) which conferred an award of £20,000 for the discovery of the North-West Passage. Despite this encouragement, no further effort —at least, none of a determined nature—to penetrate into the far north was made until 1818, when John Ross and W. E. Parry, respectively in command of the Isabella and Alexander, were commissioned by the British Government to make further search after a North-West Passage, and David Buchan and John Franklin, in command of the Dorothea and Trent, to attempt the Polar passage by way of the Greenland-Spitzbergen Sea. In the same year the act of 1776 relative to the award of £5,000 was modified into one of proportional awards (Act 58th Geo. III., chap. 20), which, by a subsequent commission, was adjusted as follows: "To the first ship, as aforesaid [belonging to any of his Majesty's subjects, or to his Majesty], that shall sail to 83° of north latitude, £1,000; to 85°, £2,000; to 87°, £3,000; to 88°, £4,000; and to 89°, as before allotted, the full reward of £5,000." *

Both expeditions of 1818, so far as the main objects to be obtained were concerned, were unsuccessful. Buchan's and Franklin's farthest north, off Spitzbergen, was 80° 34′ (July); although seemingly beyond the main body of the ice, the rapidly southward-sweeping current prevented nearer approach to the Pole.

It is a singular circumstance, as bearing upon the accomplishments of whalers and the official heads of expeditions, that the farthest north that had thus far been attained, or to which, at least, absolute credence can be given, was the northing of the two Scoresbys, who, while on a whaling voyage, on May 24th, 1806, reached the very high latitude (carefully estimated, and based upon an observation of the day before, viz.: 81° 12′ 42″) of 81° 30′. This was on Long. 19° E. of Greenwich, on the border of the great northern ice-pack, whose front had been followed from the W. S. W. over an extent of 27 degrees of longitude The younger Scoresby states regarding their position: "The margin of the ice continued to trend to the E. N. E. (true), as far as it was

London Gazette, 23d March, 1819.

visible; and, from the appearance of the atmosphere, it was clear that the sea was not incommoded by ice, between the E. N. E. and S. E. points, within thirty miles, or limited by land within 60 or even 100 miles of the place of the ship." *

One of the most determined efforts to reach the Pole, and certainly the most significant that had been made up to that time, was the one of Parry in 1827. It was the first instance where a long pedestrian traverse of the northern pack-ice was contemplated and executed as an auxiliary to the ordinary operations of the vessel of the expedition. Equipped with two open flat-floored, and runner-mounted boats, the Enterprise and Endeavor—each measuring twenty feet in length and seven feet in breadth—and a number of sledges, and thus provided for operations on both ice and water, the expedition on June 23rd left H. M. Sloop Hecla off the coast of Spitzbergen and headed due northward. No serious obstruction was encountered until Lat. 81° 12' 51" was reached, when the boats had to be hauled up on the floe-ice. From that time until the completion of the journey the course was one almost continuous struggle. The rough and hummocky ice, wholly unlike the flat plain which had been pictured by Lutwidge and Scoresby, and over which it had been assumed

---
* Account of the Arctic Regions, p. 313.

that a coach could be driven for many leagues in a direct line, with its numerous pools and water-ways, and a heavy covering of soft snow, made the work of dragging an exceedingly laborious one. Especially fatiguing was the management of the boats. The heavy equipment necessitated a constant retraverse of the same journey, so that each day's progress was a repetitionary effort made two, three, four or even five times. Indeed, Parry remarks that in some instances the traverse had to be made seven times over. The management of each boat, which, with its full complement of furniture, clothing, provisions, etc., weighed 3,753 pounds, was intrusted to twelve men, the weight per man, exclusive of four sledges weighing 26 lbs. each, being thus 268 lbs.

Parry thus describes his usual mode of proceeding:

"It was my intention to travel wholly at night, and to rest by day, there being, of course, constant daylight in these regions during the summer season. The advantages of this plan, which was occasionally deranged by circumstances, consisted first, in our avoiding the intense and oppressive glare from the snow during the time of the sun's greatest altitude, so as to prevent, in some degree, the painful inflammation in the eyes, called 'snow-blindness,' which is common in all snowy countries. We also thus enjoyed greater

warmth during the hours of rest, and had a better chance of drying our clothes; besides which, no small advantage was derived from the snow being harder at night for travelling. The only disadvantage of this plan was, that the fogs were somewhat more frequent and more thick by night than by day, though even in this respect there was less difference than might have been supposed, the temperature during the twenty-four hours undergoing but little variation. . . . . . . "

"When we rose in the evening, we commenced our day by prayers, after which we took off our fur sleeping-dresses, and put on those for travelling; the former being made of camblet, lined with raccoon-skin, and the latter of strong blue box-cloth. We made a point of always putting on the same stockings and boots for travelling in, whether they had dried during the day or not; and I believe it was only in five or six instances, at the most, that they were not either still wet or hard-frozen. This, indeed, was of no consequence, beyond the discomfort of first putting them on in this state, as they were sure to be thoroughly wet in a quarter of an hour after commencing our journey; while, on the other hand, it was of vital importance to keep dry things for sleeping in. Being 'rigged' for travelling, we breakfasted upon warm cocoa and biscuits, and after stowing the things in the boats and on the sledges, so as to secure them, as much as possible, from wet, we set off on our day's journey, and usually travelled from five to five and a half hours, then stopped an hour to dine, and again travelled four, five, or even six hours, according to circumstances. After this we halted for the night, as we called it, though it was usually early in the morning, selecting the largest surface of ice we happened to be near, for hauling the boats on, in order to avoid the danger of its breaking up by coming in contact with other masses, and also to prevent drift as much as possible. The boats were placed close alongside each other, with their sterns to the wind, the snow or wet cleared out of them, and the sails, supported by the bamboo masts and three paddles, placed over them as awnings, an entrance being left at the bow. Every man then immediately put on dry stockings and fur boots, after which we set about the necessary repairs of boats, sledges, or clothes; and, after serving the provisions for the succeeding day we went to supper. Most of the officers and men then smoked their pipes, which served to dry the boats and awnings very much, and usually raised the temperature of our lodgings 10° or 15°. . . . . . We then concluded our day with prayers, and having put on our fur-dresses, lay down to sleep with a degree of comfort, which perhaps few persons could imagine possible under such circumstances; our chief inconvenience being, that we were somewhat pinched for room, and therefore obliged to stow rather closer than was quite agreeable. The temperature, while we slept, was usually from 36° to 45°, according to the state of the external atmosphere; but on one or two occasions, in calm and warm weather, it rose as high as 60° to 66°, obliging us to throw off a part of our fur-dress. .... Our allowance of provisions for each man per day was as follows:

 Biscuit . . . . 10 ounces
 Pemmican . . . . . 9 "
 Sweetened Cocoa Powder 1 ounce, to make 1 pint.
 Rum . . . . 1 gill
 Tobacco. . . . . . . 3 ounces per week.

Our fuel consisted entirely of spirits of wine, of which two pints formed our daily allowance, the cocoa being cooked in an iron boiler, over a shallow iron lamp, with seven wicks; a simple apparatus, which answered our purpose remarkably well."

This graphic account gives little idea of the hardships, or perhaps rather, difficulties, that were encountered in the passage of the pack-sea. The rapid thawing and breaking up of the floating ice-masses, which in many instances were barely strong enough to support the weight of a single boat with its complement of men, the yielding crust and almost innumerable water-ways, and the strong southwardly-trending current combined to render progress extremely slow and irksome. Indeed, on the last day or days the southerly drift more than counterbalanced the actual advance made, so that on the 26th of July the position, as determined by a meridian altitude of the sun—$82° 40' 23''$—was actually three miles to the south of the position which had been determined at midnight of the 22nd. "Again," as Parry states, "we were but one mile to the north of our place at noon on the 21st, though we had estimated our distance made good at twenty-three miles. Thus it appeared that, for the last five days, we had been struggling against a southerly drift exceeding four miles per day."*

Parry's farthest north, a little beyond $82° 45'$, was made seemingly on the morning of July 23rd. At that time the distance from the Hecla, bearing S. 8° W., was only 172 miles, to accomplish which it is assumed fully 290 miles were covered, of which about one hundred were performed by water previous to entering the ice. "As we travelled by far the greater part of our distance on the ice three, and not unfrequently five times over, we may safely multiply the length of the road by two and a half; so that our whole distance, on a very moderate calculation, amounted to five hundred and eighty geographical, or six hundred and sixty-eight statute miles, being nearly

―――――――――――
*Narrative, p. 102.

sufficient to have reached the Pole in a direct line."

An analysis of Parry's journey leads to a simple conclusion: given a better condition of the ice for travelling, and a lighter equipment, obviating the necessity of retraverses, a decided advance could have been made in the direction of the Pole, with the possibility of even reaching the Pole. The problem presents itself: Can these conditions be met? There can be no question, as Parry himself intimates, that at an earlier season the ice, crisp and solidly frozen, would have been in a much more favorable condition for a traverse than in the months of June and July, when it becomes spongy and rotten, and harbors an almost endless series of ramifying water-lanes and pools. The constant launchings and up-haulings of the boats are an incident of an ice-journey which cannot conduce to satisfactory progress. Whether or not the great quantities of soft snow which were encountered were due to an unusually wet season, or were the product of a summer season as distinguished from the winter, cannot be precisely stated. That the summer was an unusually wet one cannot be doubted, and Parry states (p. 129) that the quantity of rain that fell was twenty times that which he had observed in any preceding summer in the Polar regions. Parry's men were not provided with snow-shoes, and it is stated that no form of snow-shoe could have been used with advantage, owing to the very hummocky condition of the ice. Not improbably, however, with a lighter equipment a decided advantage might have been derived from these articles, especially in an early season, before the forcing of the pack, when the ice presents a much less hummocky surface than at the time of its coursing and crushing.[1]

As regards the lightening of the equipment, there can be no question that, with the advantages of an experience extending over sixty years, it can be very materially effected. The journey of Mr. Peary is especially fruitful in this regard, and particularly impressive is the lesson taught by the simplicity of the personal accoutrement[2] and the food-supply. Most of all significant, however, is the fact that a long Arctic journey can be safely undertaken by but two men; and, doubtless, a small party would be preferable in a dash for the Pole to a large force whose principal function would be the hauling of its own equipment.

In concluding his narrative, Parry expresses a doubt as to the feasibility of the traverse on the lines which he attempted to carry out, and

---

[1] Lockwood and Brainard in their long journey northward were also unprovided with snow-shoes, a circumstance which was afterwards much regretted by the leader of the party.

[2] Mr. Peary's reindeer-suit, which was used on the inland ice to the exclusion, for the greater part of the journey, of both tent and sleeping-bag, weighed only 11¼ lbs.

of which he, following Franklin, was an enthusiastic advocate. He thus remarks: "That the object is of still more difficult attainment than was before supposed, even by those persons who were the best qualified to judge of it, will, I believe, appear evident from a perusal of the foregoing pages; nor can I, after much consideration and some experience of the various difficulties which belong to it, recommend any material improvement in the plan lately adopted. Among the various schemes suggested for this purpose, it has been proposed to set out from Spitzbergen, and to make a rapid journey to the northward, with sledges, or sledge-boats, drawn wholly by dogs or rein-deer; but, however feasible this plan may at first sight appear, I cannot say that our late experience of the nature of the ice which they would probably have to encounter, has been at all favourable to it" (p. 143). Parry, however, had reason to subsequently change his opinion, for in 1845, eighteen years after the accomplishment of his journey, we find him writing to Sir John Barrow as follows: "It is evident that the causes of failure in our former attempt in the year 1827 were principally two; first and chiefly, the broken, rugged and soft state of the surface of the ice over which we traveled; and, secondly, the drifting of the whole body of ice in a southerly direction. On mature reconsideration of all the circumstances attending this enterprise, I am induced to alter the opinion I gave as to its practicability in my Journal, p. 144, because I believe it to be an object of no very difficult attainment, if set about in a different manner. My plan is, to go out with a single ship to Spitzbergen, just as we did in the Hecla, but not so early in the season, the object for that year being merely to find secure winter quarters as far north as possible. . . . . . I propose that the expedition should leave the ship in the course of the month of April, when the ice would present one hard and unbroken surface, over which, as I confidently believe, it would not be difficult to make good thirty miles per day without any exposure to wet, and probably without snow-blindness. At this season, too, the ice would probably be stationary, and thus the two great difficulties which we formerly had to encounter would be entirely obviated. It might form a part of the plan to push out supplies in advance to the distance of one hundred miles, to be taken up on the way, so as to commence the journey comparatively light; and as the intention would be to complete the enterprise in the course of the month of May, before any disruption of the ice or any material softening of the surface had taken place, similar supplies might be sent out to the same distance, to meet the party on their return."

That this latter plan of Parry is a feasible one there is every reason to believe;—indeed, no other plan for reaching the Pole commends itself to equal favor, unless it be that of trying a staunch steamer in the middle or latter part of August, *in a favorable season*, and forcing a passage due northward (*vid. post.*). No expedition undertaken since Parry's time has brought in facts severely opposed to the plan outlined, and, therefore, it is not a little remarkable[1] that the effort in this direction should have received such scant encouragement, and been replaced to so great an extent as it has been by the much less promising west Greenland or American route. The fact that a shore-line could here be followed for so great a distance northward was, doubtless, the determining factor, in at least several instances, in the selection of this route—a route, the impracticability of which, so far as penetration by vessel is concerned, has been abundantly proved by the experiences of Kane (in 1853), Hayes (1860), Hall (1871), and Nares (1875).[2]

---
[1] *Vid. ant.*, pp. 22-24.

[2] Kane's brig, the Advance (120 tons), was stopped by the ice about seventeen miles beyond the entrance to Smith Sound, in Lat. 78° 45'; Cape Constitution, the furthest point reached by a sledging party (1854), is by some authorities placed in Lat. 81° 22', by others in 80° 56'. The United States of Hayes, a schooner of 130 tons, found it impossible to pass beyond Littleton Island, and winter-quarters were established at Port Foulke, in Lat. 78° 17'; a sledging party is said to have penetrated the following May to 81° 35', but most geographers

The most interesting fact connected with Parry's expedition, and one which has in a measure escaped the attention of Arctic geographers, is the exceedingly small quantity of ice that was met with at the most northerly limits reached by the expedition. Indeed, it would appear from the narrative that a staunch steamer, such as has been several times used in later expeditions, might have pushed, and without much effort, entirely through the ice which impeded Parry, and freely entered the open water beyond. Parry thus describes the conditions which presented themselves on July 25th: "So small was the ice now around us, that we were obliged to halt for the night at two, A. M., on the 25th, being upon the only piece in sight, in any direction, on which we could venture to trust the boats while we rested. Such was the ice in the latitude of 82°¾"!

This intimation of a largely open Polar sea Parry emphasizes with a concluding statement: "I may add, in conclusion, that before the middle of August, when we left the ice in our boats, a ship might have sailed to the latitude of 82°, almost without touching a piece

---

agree that the position is not to be depended upon. Hall's vessel, the Polaris, was beset in the ice, on August 30th, in Lat. 82° 16', the farthest north that had been reached by vessel up to that time, but four years later, in 1875, the powerful screw-steamer Alert, of the British Expedition, penetrated along the same route, through Robeson Channel, to 82° 27'; a sledging party extended the exploration to 83° 20'.

ESKIMOS COME TO MEET US.—CAPE YORK.

of ice; and it was the general opinion among us that, by the end of the month, it would probably have been no very difficult matter to reach the parallel of 83° above the meridian of the Seven Islands" (Narrative, p. 148). It might here be properly asked: If a sailing vessel could have penetrated without very great difficulty to the 83d parallel, how much further may not a steam-vessel have penetrated? Did Parry in fact discover an "open Polar sea"? This would appear to be the case from the statement that beyond 82° 45' no ice of any magnitude was visible; but it should be recollected that a ship's horizon, even with the advantage of a crow's nest, is an extremely limited one. From the ice-cake on which Parry rested at his "farthest north" the horizon must have been limited to some five or six miles at the utmost, and, therefore, the indications of open water need not have extended even to the 83d parallel. Indeed, Parry himself admits that a yellow ice-blink almost continuously overspread the northern horizon, a general indication of ice not being far distant.

With all allowances, however, it may be affirmed that in the summer of 1827 open water, *beyond the northern pack*, extended quite to, if not considerably above, the 83d parallel of latitude. The late Dr. Petermann was the most persistent believer in the comparative openness of the northern Polar sea, and the staunchest advocate of the Spitzbergen route as the route, for vessels, to the Pole. Commenting on the papers of this distinguished geographer, Admiral Richards, who was one of the members of the Arctic Committee of the Admiralty which recommended the Smith Sound route, states that no papers read before the Royal Geographical Society on the subject of reaching the Pole appeared to him "so sound, so logical, or so convincing." Carrying out the plan of exploration in this region, he further states: " Briefly, the proposition was that two stout and well found steamers, such as the Alert and Discovery, should seek an opening through the ice north of Spitzbergen, an attempt which has never yet been made. It is the only route which offers a prospect of success by ships, and it is impossible to deny that it does hold out a very fair prospect." (Proc. Royal Geogr. Soc., 21, p. 283).

III.

The Spitzbergen Route to the Pole.

The plan of a sledge-journey to the Pole, while it was first seriously attempted by that able officer, was not original with Parry. As has already been said, the suggestion, transmitted through Barrow, was received from Captain (afterwards Sir) John Franklin, but several years before, the younger Scoresby had already advocated this method of travel. Thus he states, in his "Account of the Arctic Regions" (1820, p. 54): "But though the access by sea be effectually intercepted, I yet imagine, notwithstanding the objections which have been urged against the scheme, that it would by no means be impossible to reach the Pole by travelling across the ice from Spitzbergen. . . . . As the journey would not exceed 1,200 miles (600 miles each way), it might be performed on sledges drawn by dogs or rein-deer, or even on foot. Foot travellers would require to draw the apparatus and the provisions necessary for the undertaking, on sledges by hand; and in this way, with good despatch, the journey would occupy at least two months; but with the assistance of dogs, it might probably be accomplished in a little less time.

With favorable winds, great advantage might be derived from sails set upon the sledges; which sails, when the travellers were at rest, would serve for the erection of tents. Small vacancies in the ice would not prevent the journey, as the sledges could be adapted so as to answer the purpose of boats; nor would the usual unevenness of the ice, or the depth or softness of the snow, be an insurmountable difficulty, as journeys of near equal length, and under similar inconveniences, have been accomplished." Scoresby's plan was, in part, a remarkable anticipation of the method so successfully followed by various Arctic explorers from McClintock to Peary, while in its special feature, the convertible sledge-boat, it foreshadows the main feature of the Ekroll project.

The relative advantages of the Spitzbergen and American routes to the Pole are set forth in a series of "discussion papers" published in the Proceedings and Journal of the Royal Geographical Society of London for the years 1865, 1866, 1868, and 1872. The participants in the discussions were Captain Sherard Osborn, Sir George Back, Admiral Collinson, Captain Maury, Clements R. Markham, Sir Leopold McClintock, Admiral Richards, J. Crawford, and Koldewey, who favored the Smith Sound route; A. Petermann, General Sabine, Sir Edward Belcher, Admiral Ommaney, Admiral Fitzroy, W. E. Hickson, Sir Roderick Murchison, Captain Inglefield, Staff-Commander Davis, and Mr. Lamont, who favored the Spitzbergen route; and Captain Allen Young, who favored both routes. The most striking fact made clear in connection with the discussions before that honorable body was the tenacity with which opinions once expressed were adhered to; so that although the controversy extended over a period of seven years—the lack of unanimity among Arctic experts defeating the application addressed to the Lords of the Admiralty in behalf of a Government Polar Expedition—there appears to have been no recession from "first positions." What is perhaps equally striking, and certainly much less explicable, is the fact, as announced by the President of the Royal Geographical Society, Sir Henry Rawlinson, that the "Council of the Society had appointed a committee of the most experienced and practical members of their body, to report their opinions upon the subject; and they were unanimously in favour of Smith Sound" (Session of April 22, 1872).

How this unanimity in favor of the Smith Sound route was obtained among the "most experienced and practical members" of the Society, in face of the contrary opinion of such men as General Sabine, Sir Edward Belcher, Admiral Ommaney, Admiral Fitzroy, Captain Inglefield, and Staff-Commander Davis, most of them tried Arctic or Antarctic explorers, it is difficult to conceive. The later experiences of Hall, Nares, and Greely, added to the experiences of all previous explorers in the same region, only more clearly demonstrate how ill-judged must have been the facts which were marshalled up in favor of the views of the "special" committee.*

The more remarkable does the opposition to the Spitzbergen route appear in view of the fact that prior to 1865 not a single attempt had been made by steam-vessel to penetrate the northern pack at a high latitude, and between 1865 and 1872 Nordenskjöld's effort in the Sofia—undertaken more especially for the purpose of determining the relative advan-

---

* Since writing the above, I have found in a paper by Mr. Markham, on the Arctic Expedition of 1875-76 (Proc. Royal Geogr. Soc., 21, p. 540), that in the report of the Committee above referred to, "all mention of reaching the North Pole as an object, was purposely excluded." The unanimity on the part of the Committee thereby becomes much more intelligible, although it looks somewhat as though the change of base was intended to deceive the Government, the previous discussions clearly indicating the desirability of making for the Pole. Significant in this connection is the instruction given by the Arctic Committee of the Admiralty (Admirals Richards, McClintock, and Osborn) that "the scope and primary object of the Expedition should be to attain the highest northern latitude, and, if possible, to reach the North Pole." The members constituting the special Committee of the Royal Geographical Society were Sir George Back, Admiral Collinson, Admiral Ommaney, Admiral Richards, Sir Leopold McClintock, Captain Sherard Osborn, Dr. Rae, Mr. Findlay, and Clements R. Markham.

tage of an autumn navigation over that of the summer or spring season—when the latitude of 81° 42′ (Long. 17½° E. of Greenwich) was attained, was the only one that could in any way be considered in the nature of a real attempt. Nordenskjöld reached his highest point in the latter half of September,* and was stopped by a sea densely packed with broken ice. He concludes from his experiences that: "The idea itself of an open Polar Sea is evidently a mere hypothesis, destitute of all foundation in the experience which has already by very considerable sacrifices been gained; and the only way to approach the Pole, which can be attempted with any probability of success, is that proposed by the most celebrated Arctic authorities of England, viz: that of—after having passed the winter at the Seven Islands, or at Smith Sound—continuing the journey towards the North on sledges in the spring." It is a little difficult to harmonize the first part of this conclusion with Nordenskjöld's own belief, as expressed by himself, that if the "year had not been unusually unfavorable with regard to the condition of the ice, we might in all probability have proceeded a considerable distance farther, perhaps beyond 83° N. Lat." If this is

---

* "Probably the highest northern latitude a ship has ever yet attained." Nordenskjöld, Swedish North Polar Expedition of 1868. Journ. Royal Geogr. Soc., London, XXXIX, p. 142, 1869.

assumed to be easily possible—in other words, if a favorable season would permit of an advance farther northward of some 90–100 miles—on what foundation rests the assurance that another hundred, two-hundred, or even five-hundred miles might not have been covered in the same way? Buchan (with Franklin) and Parry, the only explorers, recognized as such, who broke through or penetrated beyond the outer ice-barrier, found at their terminals open water—Parry at a point only seventy statute miles beyond Nordenskjöld's "farthest"—and this form of testimony must be accorded weight beyond anything that is derived mainly from supposition.* To this testimony must be added

---

* The statement (*vid. ant.*) that no serious effort to penetrate the pack-sea north of Spitzbergen had been made by steam-vessel up to 1872 can be extended to our own day. Petermann, commenting on the lack of effort made in this direction, and on the successful penetration of Captain (Sir George) Nares to Lat. 82° 27', affirms his belief that a properly equipped expedition would have no more difficulty in steaming through the Spitzbergen Sea ice than Ross had of *sailing* through the Antarctic ice thirty years before. With the effort of 1875-76 to penetrate the northern ice by way of Smith Sound and Robeson Channel applied to the Parry route, it is thought certain that the Pole would have been attained (*Petermann's Mittheilungen*, 1877, p. 24). Nordenskjöld's positive statement regarding the non-existence of an open Polar Sea is in marked contrast to the cautious opinion expressed by Weyprecht, the virtual Commander of the Austro-Hungarian Expedition of 1872-74, to Petermann, under date of November 1, 1874 : *Erstens sind die Schlüsse auf ofenes Polarmeer in höchsten Norden eben so falsch wie diejenigen der absoluten Undurchdringlichkeit des vor dem neuen Lande vorliegen-*

the corroborative data supplied by the early
whalers, which, while they may not be considered to be of an absolutely satisfactory charac-
ter, have certain elements of plausibility about
them. Indeed, it would seem on trustworthy
evidence, as adduced by the late Dr. Petermann,
that a whaler, the Truelove, of Hull, as late
as 1837, penetrated without much hindrance
to 82° 30′ (Long. 12°–15° E.), and so experienced a whaler and Arctic navigator as Captain David Gray, of Peterhead, affirmed his belief that in 1874, having penetrated to 79° 45′,
beyond which little or no ice was to be found,
he "could have gone up to the Pole, or at any

den Eises (*Mittheilungen*, 1874, p. 452). The failure of the
Swedish Expedition of 1872-73 (the main object of which was
a Polar sledge-journey) to reach a high northern latitude has
little bearing upon the question at issue, and still less, the
efforts of Koldewey, the Commander of the Second German
Expedition of 1869-70, to penetrate the east Greenland ice-
barrier between the parallels of 75° and 76°. The main purpose of the Austro-Hungarian Expedition of 1872-74, under
Weyprecht and Payer, was the forcing of the North-East Passage; the remarkable drift of the Tegethoff, through which
Francis-Joseph Land was accidentally discovered, deals with
an ice-formation very different from that of the north Spitzbergen sea. Lieutenant Payer thus expresses himself in relation to the Polar problem: "Our own track to the north of
Novaya Zemlya carries no weight in considering this question, for we were indebted for our progress to a floe of ice,
and not to our own exertions." The Dutch Expedition of
1878 was in the nature of a reconnaissance—a preliminary to
more extended operations in the future; Captain de Bruyne,
in the schooner William Barents (79 tons), reached the front
edge of the ice north of Spitzbergen in Lat. 80° 18′.

rate far beyond where anyone had ever been before." *

The obstacle that has, perhaps, stood most in the way of the assumption of the Spitzbergen route to the Pole, by English and American explorers at any rate, is the so-called "canon" of Arctic exploration: "Stick to the land-floe" or "follow a line of coast". Among the more urgent and persistent upholders of this doctrine have been the late Admiral Sherard Osborn and Mr. Clements R. Markham. The maxim is in some respects, doubtless, a good one, but it is questionable if the experience which has thus far been acquired from Arctic ventures is sufficient to warrant its being pinned to the mast, for all time. Indeed, our experience with the open pack is still much too limited to permit us to postulate a measure of the possibilities which it

---

* Letter to Mr. Leigh Smith (Proc. Royal Geogr. Soc., 19, p. 179). From a passage contained in Witsen's "Tartarye," published in 1707, Petermann infers that Francis-Joseph Land may have been discovered and reached three hundred years before the arrival there of the Tegethoff. It recites that a certain Captain Cornelis Roule had "been in $84\frac{1}{2}°$ or $85°$ N. Lat. in the longitude of Novaya Zemlya, and has sailed about forty miles between broken land, seeing large open water behind it. . . . . . He found lots of birds there, and very tame." Petermann, commenting on this statement, states that the longitude of Nova Zembla passes right through Austria Sound and Francis-Joseph Land; that the latter is a "broken land," behind which Lieutenant Payer saw "large open water," and "found lots of birds." The coincidental description is certainly striking.

presents; and at the present time, with our intimate knowledge of past exploration, the Nansen expedition is planned wholly on the basis of open pack service. The failure, by the heavy vessels of the British Expedition, to penetrate the ice north of Robeson Channel, and the determination that the northern extremity of Greenland, or of its outlying islands, probably terminates not far from the 84th parallel of latitude, practically dispose, for the present, at least, of the Smith Sound route as an available route to the Pole. Sir George Nares himself states his belief that "great difficulty will be found in advancing much nearer to the Pole by the Smith Sound route than has already been attained, either in a ship, or by boat, or sledges, unless, indeed, the coast of Greenland—contrary to my expectations—trends to the northward beyond Lat. 83° 20′ N."* But if this route is impracticable, manifestly the only resource is to the open pack—unless, indeed, a coast-line be found to extend northward for a considerable distance beyond the farthest point to which Francis-Joseph Land is known to extend—and it becomes merely a question as to which pack offers the greatest advantages for a direct traverse. And no special regret need be expressed for this one alternative, for, despite all efforts that have been made to magnify the

---

* Proc. Royal Geogr. Soc., 21, p. 281.

success of recent expeditions, and to minimize the results obtained by earlier explorers, impartial critics must admit that of all attempts made to reach the highest latitude, the one of Parry, in 1827, taking the essential circumstances into account, was by far the most successful and remarkable; and this expedition was conducted on the open pack. That in those early days of Arctic exploration, with the rude appliances of food and equipment which were then available, he should have been able to penetrate to within forty miles of the farthest north that has since been made possible —made possible with an experience extending over fifty years, and an expenditure of money, the vastness of which, as a vehicle of Arctic exploration, could never have been even remotely contemplated by Parry—is a fact which appeals strongly in favor of the open pack. This appeal comes the more forcibly since, as has already been stated, no sincere attempt on the lines laid down by Parry has been attempted since his time, and consequently no contradiction to the belief in the easy feasibility of this route given.

The question of success, in fact, seems to resolve itself entirely into the possibilities of sledging, and Parry has expressed himself decidedly on this point. He believed that in a sufficiently early season, before the breaking up of the ice, he could accomplish thirty miles

CAPE YORK ESKIMOS.—GREENLANDERS.
(Arctic Highlanders.)

per day. Assuming that only one-half of this amount were made good, which would approximately represent the rate of travel of Lockwood and Brainard in a twenty-two days sledging trip from Cape Bryant to Lockwood Island and return, the journey, from the southern border of the ice-pack to the Pole itself, could be accomplished in about forty days, or less. Parry's failure to reach a further northing than 82° 45' was due, as has already been explained, to the southern drift which more than antagonized the exceedingly slow progress in the contrary direction which the heavy equipment permitted. There is reason to believe that had Parry had the advantages of the possibilities of compacting and of lightening in construction which years of experience have brought forth, the Pole itself might have been attained. This is the view of so experienced an Arctic explorer as Sir Leopold McClintock, who, in discussing the methods and merits of "Arctic Sledge-travelling," states his conviction that "the failure of Parry's attempt to reach the North Pole in 1827 was largely due to the great weight of his boats, and the consequent difficulty of dragging them over the ice. This error we have attempted to correct by supplying boats* of considerably less than half the weight of Parry's."

---
*To the British Polar Expedition of 1875-76. Proc. Royal Geogr. Soc., 19, p. 475.

One of the numerous fallacies connected with Arctic exploration is the supposition, held by some, that sledge-journeys cannot be prosecuted for any very great distance over the pack, or indeed anywhere. It is true, as stated by Admiral Richards (Proc. Royal Geogr. Soc., 21, p. 283), that the "longest distance ever accomplished by any one sledge-party, or by any combination of sledges, in one direction did not exceed 360 geographical miles in a straight line," but this fact, in itself, barely warrants the statement that "sledge-travelling with a view to reaching the Pole was at an end forever." This position is justly combatted by Dr. Rae, who holds that it is based upon the experience of but a single party, and, that it is "still quite possible that sledge-journeys might be made to the North under more favorable circumstances." We have as yet barely any experience bearing upon the extreme possibilities of sledge-journeyings, and what there is points much more nearly to the positive rather than to the negative side of the question. Indeed, there would seem to be no limitations to the work of a properly equipped and conducted sledge-party, whether on the open pack or elsewhere, and Sir Leopold McClintock probably justly states that "there is now no known position, however remote, that a well-equipped [sledging] crew could not effect their escape from, by their own unaided effort."

Payer, the commander of the land journeys of the Austro-Hungarian Expedition, affirms his belief that the only chance of traversing the Polar realm is by means of sledges (communication addressed to Petermann, under date of November 5th, 1874). Just as firmly as he was wedded to the proposition of penetrating the Spitzbergen Sea by means of powerful steam-vessels, so was Petermann opposed to the principle of long sledge-journeys, in the ultimate success of which he had no faith. A strong support was claimed to be given to his views by the failure of the Swedish Expedition of 1872-73, when Nordenskjöld and Palander made an ineffectual attempt to traverse the ice north of Mossel Bay, Spitzbergen. In this experiment, conducted over an exceedingly heavy and hummocky ice, progress was on some days measured by less than half a mile. A weight of 280 pounds was distributed to each man of the expedition, or very nearly that which was allotted to the members of Parry's party.* The ill-success attending this expedition must be given due weight. It is probable, however, that men more accustomed to sledge-journeys than were the members of the Swedish Expedition, or endowed with a firmer determination, would have accomplished more. The heavy equipment seems again to have been the principal bar to progress, for

---

* Palander, in Petermann's Mittheilungen, 1873, p. 348.

Nordenskjöld admits, that with the assistance of forty reindeer, in place of the single animal which was at the service of the party, even with the exceptionally unfavorable condition of the ice, they would have been able to penetrate very far (*sehr weit*) beyond the Parry Islands.*

Brought to its simplest expression the exploration by way of the Spitzbergen Sea means:
1. A direct traverse of the sea and pack by staunch steam-vessels, or
2. The accomplishment of a large, and probably the greater, part of the journey by means of sledges and portable boats.

The probability seems to lie with the side of the alternative, and in every event its contingencies must be provided for. It cannot, however, be too strongly insisted upon that preliminary reconnaissances to the main effort, regulated to meet the exigencies of such effort, should favorable conditions present themselves, are an essential to success. There is no reason why an expedition fitted out to make a direct passage northward should, in the event of specially unfavorable conditions presenting themselves, squander its energies in an attempt to accomplish a forlorn hope, and return with a miserable record of failure. It would be far better to have it return than expose it to the chances of disaster, and better to outlive the

---
*Petermann, 1873, p. 445.

loss of glory than suffer the penalty of misdirected pride. The ready accessibility of the Spitzbergen Sea places it within easy reach of expeditions of only moderate cost, and there ought to be little difficulty in organizing yearly voyages to the north with the hope of once stumbling upon a specially favorable season.

As to just which portion of the Spitzbergen Sea is most available—whether on the line followed by Parry or eastward by the shores of Francis-Joseph Land (the route favored by Greely, Markham and Melville)can in our present knowledge scarcely be predicated. The western shores of Francis-Joseph Land, should they be found to extend much to the northward of the limitations which have been generally assigned to them, would doubtless possess all the advantages which are involved in the Arctic canon of keeping by the land; and the fact that the land has been traced (optically) quite to the 83d parallel is a circumstance which must be given weight in the consideration of the possibilities of disaster and the necessities of a forced retreat. If winter-quarters are to be established on either Spitzbergen or Francis-Joseph Land, preliminary to an early spring sledge-journey, then manifestly Francis-Joseph Land has the advantage of distance, since its northern apex is nearer to the Pole by at least one hundred and thirty miles than is the northern extremity of Spitzbergen. The western or "lee" exposure of the ice-foot of Francis-Joseph Land, moreover, offers the proper requirements for an even surface, and not improbably, also, in this higher latitude, nearer to a centre of currental motion, the pack itself would be found to be more nearly stationary than at a position further south.

The chances for a successful issue to a sledge-journey northward from Francis-Joseph Land are seemingly the best which any region offers; and whether the Pole is itself attainable by this route or not, it is all but certain that a much higher latitude would be gained (and gained with comparatively little effort) than has ever before been possible. With winter-quarters established between the 82d and the 83d parallels of latitude, or whatever the farthest point might be to which the impedimenta of an expedition could be carried by steamer, a northern start could be begun with the month of March—perhaps still earlier—and a return effected in time, possibly, for a final departure from the region still the same year. At least five months would thus be available for the purposes of sledging, and it is probably not too much to expect that from five to ten miles, on an average, could be covered every day of this period. With the lowest amount here stated, of five miles, the seventy-five days outward journey would conduct to the very gates of the Pole, or to within sight of the immediate Polar tract if any high land existed within the boundaries. Should it be desired to spend two winters in the region, then, manifestly, the length of time of exploration could be considerably extended. The most serious contingency to be guarded against on an expedition of this nature would be that of driftage arising from the breaking and dissolution of the pack before the return of the exploring party, an event that would almost certainly take place full two months before the completion of the enterprise. This condition would be in measure met by the relief of portable boats, but there is no reason why the supporting vessel of the expedition should not cruise through the broken pack-sea along its open front and chance the meeting with the returning parties. Such contemplated assistance would probably not be absolutely necessary, nor would it be assured of being carried out, but the attempt to realize it would cost little beyond effort. The advantages of the Francis-Joseph Land route to steam-penetration are less apparent, since seemingly the 82d or 83d parallel can be reached without much effort, and, moreover, it may be questioned whether the ice accumulation about the newly discovered land is not in fact heavier than farther westward in the more open oceanic tracts.

THE EAST GREENLAND SEA.

What facilities for obtaining a high northing

are afforded by the east Greenland Sea,—or more properly, the sea that is included between the eastern shores of Greenland (above the 78th parallel of latitude) and the outlying border pack—is not known; nor, in fact, is it positively known that such a sea exists at all as a permanency, although from the data obtained by numerous whalers there can be no question of its existence at times. Petermann was an urgent advocate of the possibilities of this route, but he had manifestly much underrated the difficulties and dangers attending the passage of the outlying pack-ice. The recent failures of the Danish Commander Ryder to penetrate this ice add testimony to the unfortunate experiences of the Second German Expedition under Koldewey in 1869-70. Yet it is by no means certain that in a specially propitious season good fortune would not attend an expedition in this direction. Captain David Gray had apparently no difficulty in forcing the ice in 1874, and, as is well known, an open sea in 1817 and 1818 washed the east Greenland coast as far north, seemingly, as the 79th parallel. The recent explorations of Mr. Peary, by demonstrating that a position on the northeast coast, near to the 82d parallel, can be obtained through an overland journey from the west, opens up a new possibility, and even good promise for this route. Starting from this point (which could probably be reached from

winter-quarters established on Inglefield Gulf as early as the month of May, or even the middle of April) as a base, the Parry method might be essayed thence over the open frozen sea. This route would probably have an advantage over that pursued by Parry by being placed (presumably) beyond the zone or belt of greatest driftage; on the other hand, the preliminary traverse of some four hundred miles of land, with a full equipment, is a disadvantage which cannot as yet be fully measured in the scale of possibilities. That this traverse is not a decided obstacle to a protracted excursion beyond, is indicated by the preparations which are at the present time being made by Mr. Peary for the examination of the eastern and northern coast-lines of Greenland *through an expedition starting from the west.* We know as yet nothing of the condition of the sea-ice north or northeast of Greenland; seen from Independence Bay, in Lat. 81° 37′, it appeared to Mr. Peary to be largely destitute of hummocks, but the distance of vision was such as to make the determination somewhat doubtful. At the time of his visit, the early days of July, no water was visible, but this fact is no evidence of permanent gelation. Melville Bay is at the same period of the year very largely solid, although two or three weeks later little ice is to be seen.

If Greenland be used as a base of operations

to reach the Pole, then the Smith Sound route commends itself to better favor than the overland traverse. It is almost beyond question that a staunch vessel could during one or two months of every year penetrate to at least the 81st parallel, and find a safe anchorage in one of the numerous deep bights or fjords which open into the western waters. An advance post for a sledging party could probably be established the same season considerably beyond the "farthest" of Lockwood and Brainard, or within some four hundred miles of the Pole, whence the main journey over the frozen sea—unless, indeed, the island-masses lying north of Greenland extend much beyond their assumed limits, and permit of a coastal traverse—would be continued in early spring. The southwestwardly trending swing of the Polar ice, and the absence of sufficiently extensive channels or avenues of discharge in this region, make it highly probable, however, that the open pack would be in a much more broken and unfavorable condition for travelling over than in the region about Francis-Joseph Land, where to the westward a broad avenue of release is afforded by the Greenland-Spitzbergen Sea. In its several relations the Francis-Joseph Land route holds out the greatest promise to a boat-and-sledge journey.

Captain Albert H. Markham has in his work, "A Polar Reconnaissance" (1881), strongly advocated the real advantages of the Francis-Joseph Land route, and the opinion of no Arctic explorer is worthy of higher consideration than that of the intrepid commander of the "farthest north" sledging party of 1875-76. His conclusion is stated as follows: "From a careful study of all that has been achieved in the far North, I am more than ever convinced that a greater amount of success will be gained by the exploration of the region in the vicinity of Franz Josef Land than in any other part of the Arctic regions" (p. 319). In the Preface to this work, Mr. Clements R. Markham, for many years the unbending advocate of the Smith Sound route, gives in his adhesion to the views of Captain Markham. "Here, therefore, is the route for future polar discovery. Here an advanced base may be established within the unknown region, whence scientific results of the utmost interest will be secured: and here the nearest approach to the North Pole can be made."

NOTE. Since the foregoing was sent to press announcement has been received that an attempt to reach the Pole by way of Francis-Joseph Land will be made this summer by Mr. Frederick G. Jackson, a Fellow of the Royal Geographical Society of London. The plan of journey, as outlined by Mr. Jackson in the London *Times*, is as follows:

"Franz Joseph Land, according to Sir Allen

Young and other authorities, would be easily accessible any ordinary year, and it is the only point where one could establish a base whence to extend a secure line of march northward, with a possibility of safe retreat, if necessary, to one's headquarters. I propose, therefore, to avail myself of this fortunate fact, and sail this summer, if possible, for the southern coast of Franz Josef Land. I shall hope to arrive there in time to make a rapid reconnoitre northward (and perhaps reach a point further north than the Austrian limit) before moving into winter quarters, In the following year I have every expectation of being able to push forward with a considerable quantity of stores and establish a second depot, probably in or about the 84th or 85th degree. Having there established a base in this high latitude, I shall have ample time, I believe, to make a third march northward (should there be land beyond this limit). If there be land, this third march should enable me to establish a third depot, within 200 miles of the Pole, and here I should winter. I should then have the whole of the next summer in which to seize the opportunity to make a push to the ultimate object of the expedition. If there be not land northward of the second depot (84 degrees) I should hope to firmly establish myself there for the winter, and prepare to make the necessary forward advance in the next spring. But for all that is known to the contrary, Franz Josef Land may extend even to the Pole. Should this be the case, one summer with fair weather may prove sufficient. I propose to attempt this with a small party of not more than ten men. We should have light equipment, sledges, dogs, etc. By establishing a chain of depots, I may point out that we should escape the burden of carrying with us a large quantity of stores and provisions at the very time when we should wish to move most rapidly."

The results of this venture will certainly be awaited with much interest. If carried out with a strict observance of the necessities and possibilities which the modern Arctic study has developed, and under the force of that patient energy which so triumphantly bore the British pennant across the ice-fields in 1875, it cannot easily fail to plant an important stepping stone in the path of Polar exploration. To it may we confidently look for much new light upon the Arctic question, if not for the absolute resolution of the problem itself.

A GREENLAND OFFICIAL'S WIFE.
(Mrs. Sofia Baumann)

IV.

THE PEARY RELIEF EXPEDITION.

On June 6, 1891, the good ship Kite, a barkentine whaler of the old type, and measuring barely forty yards in length, lay alongside one of the busy Brooklyn wharves, eagerly scanned by hundreds of eyes for the little that distinguished her from the neighboring craft. Neatness or cleanliness was not a characteristic of the vessel, for she still bore traces of seal-strife and struggles among the ice of Newfoundland's coast.

To certain peculiarities of structure was added a suggestion of the odor of oil and blubber, and if these were not in themselves sufficient to indicate the rank of the vessel, it could readily have been told from the iron bow-cap, and that singular aerial castle known as the crow's nest. However insignificant and humble the Kite may have appeared beneath the tall hulls and masts that surrounded her, she bore a trim side to the waters of an open sea, and in her adopted port of St. John's she is a craft with a history and a name.

Prior to the date above mentioned, the most distinguished name associated with the vessel was that of her then master, Captain Richard

Pike, a sea-dog devoid of those characteristics which entitle one to the designation of "bluff," but who, despite this deficiency, had already, on two occasions, done service among the ice-fields of the far north. To his hands, as ice-master, the Government in 1881 entrusted the fate of the Proteus—the ship which conveyed the Greely party to their point of location, near the eighty-second parallel, which was destined to serve as a home of desolation for a period of three years.

In 1883, on the organization of the second Greely Relief Expedition, under Lieutenant Garlington, Pike was again pressed into Arctic service as the ice-master of the relief-ship Proteus, the crushing of which among the ice-floes of Smith Sound, off Cape Sabine, has become a matter of history. The ten years that have elapsed since the day of the disaster have not yet sufficed to wipe off the cloud from the genial tar's brow, over which the shadows of fifty-three years have now gathered. A quiet resolve never again to enter the Arctic seas was brushed aside when, in 1891, the Kite was chartered to convey the expedition of the Philadelphia Academy of Natural Sciences to the Greenland waters, and a demand made for the services of an experienced ice-master and pilot.

The Kite left her anchorage among the Brooklyn hulks on the afternoon of June 6th, carrying as her passenger list the members of the Peary party—Civil-Engineer Robert E. Peary, U. S. N., Josephine Diebitsch-Peary, Dr. F. A. Cook, Langdon Gibson, Eiwind Astrup, John T. Verhoeff, and Matthew Henson—and an auxiliary body of "summer" investigators, to which the writer had the advantage to be attached. After varying incidents of one form or another, the good little craft put in at Godhavn, the capital of the Northern Inspectorate of Greenland, on June 27th, and on July 2d, almost exactly opposite the Devil's Thumb, buried her nose in the pack-ice of Melville Bay, from which she was destined not to emerge until three weeks later.

It was during the traverse of this ice, on July 11th, that Lieutenant Peary met with that mishap—the breaking of the lower right leg, which came near to shattering the enterprise upon which the commander had for years set his mind. In a constitution less vigorous, and a mind less heroic, such an accident would have annihilated all aspirations for success, even in the most favored undertaking; but to Mr. Peary and his gallant wife, it was but an incident, the passage of which was to be determined only by future events. On July 24th, the Kite reached McCormick Bay, on the southern shores of which, and in the shadows of the bright-red cliffs which make up much of what belongs to Cape Cleveland, the Peary winter-quarters were established. Many pleasant memories attach to the little retreat beneath the boards and tar-papers of the Redcliffe House, where probably was passed the most comfortable and homelike winter in the far north which it has been the lot of Arctic explorers to experience. On July 30th, the Kite, with the auxiliary party aboard, steamed out of McCormick Bay, leaving the North Greenland Expedition to shift for itself during the many months which were to follow before contact with civilization could again be made possible. It was during these months, extending from August to May, that those careful studies of possibilities were made which have rendered practicable the most remarkable ice-journey that has ever been undertaken, and brought to the geographer the solution of one of the few significant problems which remained open to him. Greenland has been demonstrated to be an island, whose general northern contours lie south of the eighty-third parallel.

Probably no scientific expedition originating on this side of the Atlantic has attracted more general attention than the one which Mr. Peary has but recently brought to a successful termination. Its special feature, the traverse in a due geographical course of nearly four hundred miles of the inland ice, was the pivot about which much of this interest centred. The bold manner in which the expedition had been conceived, involving an almost total departure from the methods that had been followed by all previous expeditions to the far north, and the circumstance that the party of exploration had been reduced to less than a handful of men, lent additional interest to the enterprise. To the scientist the interest was more than a purely sentimental one. The successful issue of the expedition meant the solution of some of the most perplexing problems which were yet open to the investigator. The conditions which determined the limitation of man's habitation on the globe, the nature and extent of the great Greenland ice-cap, and its relation to the ice accumulation of the Glacial Period, and the distribution of plants and animal forms beyond the boundaries of the ice-cap itself, were the topics of special scientific interest which linked themselves with the main geographical inquiry—the determination of Greenland's northernmost boundaries.

The only weak point of the Peary Expedition was the failure to make adequate provision for a return to civilization after the accomplishment of the inland journey. It was the intention of the leader to make his way leisurely down the coast in open whale-boats—two of which had been specially constructed for the purpose—and dare the ice and storms of Melville Bay as he had dared the winds and snows of the inland ice, from the sea-level to 8,000 feet elevation. Once across the Bay, the journey could be readily continued to Upernivik or Godhavn. The passage in open boats of Melville Bay has been accomplished, either in whole or in part, on several occasions—by Kane, in 1855; by Bessels and Buddington, in their retreat from the Polaris, in 1873; by Pike and Garlington, in their retreat from the Proteus, in 1883—but always with great difficulty, and under the guidance of an ample force of able-bodied men. In the present instance, the party, including the courageous wife of the commander, numbered but seven members, too limited in strength, probably, to undertake the risks which the journey entailed. Under the circumstances it seemed eminently proper that assistance be rendered to the returning party, and it was with a just appreciation of this position that the Philadelphia Academy of Natural Sciences undertook the organization of a Relief Expedition.

Under my command, as leader of the Expedition, were associated Henry G. Bryant, the successful explorer of the Grand Falls of Labrador, second in command; Dr. Jackson M. Mills, surgeon; William E. Meehan, botanist; Charles E. Hite, zoological preparator; Samuel J. Entrikin; Frank W. Stokes, artist; and Albert White Vorse, most of whom had already been tried in mountain or camp work of a more or less arduous nature. The Kite was again chartered as the vessel of the expedition, and with her, the tried captain of the Proteus, Richard Pike. The possibilities of the Relief Expedition were such that no anticipatory plan of action, except as it was indicated in its broadest details, could be determined upon as a finality. The contingencies that might present themselves were too numerous to permit of simple resolution, and therefore full scope was given the expedition to meet the exigencies of the moment. It was, however, considered a necessity to pass Melville Bay at the earliest possible time consistent with an assurable amount of safety to the vessel, as once beyond the ice and waters of that much-dreaded section of the Arctic world the passage to McCormick Bay could be made without hindrance of any kind. The experience that has been brought down from the various Arctic expeditions, and more particularly from the different whalers which every year traverse much of the northern icy seas, has infused an element of certainty into Arctic navigation which could hardly have been realized by the heroes of a period twenty-five or thirty years ago. The capture, by the Melville Bay pack, of McClintock's Fox in the latter part of August, 1857, could scarcely be paralleled to-day, except as the outcome of ignorance or disregard of every-day knowledge. In an average season Melville Bay can be traversed about as readily as almost any large body of water lying southward, while its earliest seasonal passage can be predicated with a precision almost akin to mathematical calculation. The hard pack-ice which has accumulated as the result of the winter's frost, and has to an extent been held together through the large bergs which are here and there scattered through it, usually shows the first signs of weakness between July 10th and 20th. Large cakes or pans of ice have by that time succumbed to the powerful oceanic currents that are directed against them, and detaching themselves from the parent mass, float off to find new havens of their own. The weakening process continues until most of the ice has been either removed or melted away, and before the close of the fourth week of July little beyond shore-ice (shore-pan) remains to indicate the barrier which but a few days before rendered a passage all but impracticable. The trend of the ice is northwestward through the Bay, then westward to the American side, and finally south to the open sea. It was the purpose of the Relief Expedition to reach the southern boundary of the Melville Bay pack on or about the 20th of the month, and there watch the movements of the ice until the opportunity for action arrived. An earlier traverse might possibly have been made through persistent "butting" of the ice, but the dangers incident to this form of navigation were such as to render slowness a prudent measure of safety.

At 2.30 of the afternoon of July 5th the hiss of the siren announced to the loiterers on the wharves of Newfoundland's capital that the Kite was about to depart on her second voyage to the Arctic seas. A few moments later the vessel swung from her wharf, and amid a chorus of hurrahs and the shrill accompaniments of steam-whistles, started on her mission of good-will northward. The bold sandstone cliffs guarding the entrance to St. John's Harbor, aglow with the warm sunshine of a "typical" day, were soon dropped in the rear, albeit the rate of travel was somewhat less than seven knots an hour. Few of the St. John's sealers are rated for more than nine or ten knots; of the entire fleet the Kite is about the least swift, but what she lacks in this regard is more than compensated for by a staunchness of construction and a commodiousness of design which render her specially adapted for the purpose for which she was selected. The first few days of the voyage were wholly uneventful, and almost without incident. In the afternoon of the 10th, after heavy fogs had largely obscured our course, suspicious cakes of ice indicated a near approach to the Greenland coast. At midnight of the 11th, when a rift in the fog first revealed the presence of Greenland's serrated mountains, the guard-rails of the vessel were almost overtopped by the ice; fortunately the pans were not sufficiently packed to cause serious alarm for our position, despite the disagreeable feature which the presence of an ever-falling fog added.

The point of the Greenland coast opposite to our position was approximately the great Frederikshaab glacier, one of the most gigantic of the almost endless number of ice-sheets which radiate off from the inland ice to or toward the sea. In the passage of this portion of the coast the summer previous no sea-ice beyond freely floating bergs was encountered, but in the present year the ice extended fully seventy miles farther northward, and, as subsequent events showed, it was the heaviest accumulation that had been known for several decades. The southern ports of Greenland had for weeks been inaccessible, while the vessels of the cryolite fleet, for two months or more, had found scant quarters amid the jam that was impending. Wreckage appeared in scattered masses, and intelligence of disaster turned up everywhere. The Kite finally extricated herself shortly before noon of the 12th, when about opposite Lichtenfels, the northernmost point which the lower or Cape Farewell ice is known to attain.

Fog and rain followed the expedition for another thirty-six hours, but on the morning of the 14th day broke with a splendor and luminosity unknown to regions outside of the Arctic Circle. The Greenland coast loomed up brilliant for a length of a hundred miles or more, its rugged mountain peaks, here and there flecked by the snows of lingering winter, or forever shrouded in the white mantle of a perpetual ice-cap, forming a continuous panorama not unlike what is presented to the observer from the lower mountain summits of Switzerland. It is true that the loftiest peaks are here but four or five thousand feet in elevation, but the absence of foreground and the low descent of the snow-line combine to produce an exaggerated optical effect which is most delusory, a deception that is only further strengthened by the *Hörner* and *aiguilles* which everywhere recall the Alps. It is Switzerland in miniature, with a smooth, glassy sea to receive the reflections which in old Helvetia bathe in the waters of her deep blue lakes. Seventy miles to the northward a slight heaving of the horizon indicated the position of the basaltic cliffs of Disko Island, under the lee of which are nestled the few huts and houses which together constitute the capital of the Northern Inspectorate of Greenland, Godhavn or Lievely. The average mind which conceives of a journey to the far north as being one of only hardships and terror, finds it difficult to realize that this is the "land beyond the Arctic Circle;" the warm sunshine, the placid sea, and the absence, except in scattered flecks, of those impending bergs which have fastened themselves as time-honored necessities upon the eye of the imagination, fail to do justice to the modern conception of the Arctic world. The temperature at 8 A. M. was 45° F., but at noon it had risen to 50° F., and in the sun the station of the mercury among the seventies did away with all thoughts concerning wraps and heavy underwear.

At 5.30 in the afternoon we arrived off Godhavn, and shortly afterward passed through the formality of taking on a pilot—an Eskimo of unmistakably European lineage. Swarthy Frederick, the interpreter to the British Polar Expedition of 1875-76, and the associate of Peary in 1886, was among the first to greet us, bringing with him a number of his tribe, young and old—but all males, as no females are permitted to board the incoming vessels—prepared to partake of a lasting hospitality of the ship's steward, and to effect such barter as would yield to them the advantage of a few krones or of a shirt or pair of pantaloons. The latter article was prized beyond measure, but its acceptance was dependent wholly upon a proved freedom from holes and patches. Danish sovereignty has long since infused a civilized aspect into the costume of the southern Eskimos, and hence the demand for articles which would be scorned by most of their brethren of

EARLY MORNING IN THE NORTH WATER.
(Petowik Glacier)

the north; European trousers and a blue cotton outer shirt or *anorak* now take the place, as a summer attire, of the seal garments which were a necessity in the antecedent periods of barbaric existence. Among those who had come out with the first boat-load of visitors to the Kite was an old Eskimo who had in 1870 conducted Nordenskjöld to the famous "meteoritic" region of the Blaaberg, on Disko Island, whence were obtained the large blocks of native iron, commonly known as the iron of Ovifak or Uifak, concerning the origin of which, whether meteoric or telluric, so much has been written and argued by geologists and mineralogists. I was at the identical locality with the same Eskimo in the summer of 1891, and fortune threw in our path a stone of some two hundred and seventy pounds weight, for which a reward of £5 was given. Suspecting that there might be a return expedition this year, the Eskimos had shrewdly made a further examination of the desolate spot, with the result of finding a number of additional blocks of the desired material; these had been carefully placed to one side awaiting my return, and were now placed at my disposal, together with much other geological material that it was thought I might be interested in.

Our purpose in putting in at Godhavn was primarily the presentation of official credentials from the Danish Government, and the obtaining of certain effects which were considered desirable for the expedition. Godhavn, or, as it is commonly known to geographers, Disko, as the capital of the Northern Inspectorate of Greenland, is the official seat of one of the two highest dignitaries of the land, the Inspector. Of a population counting less than one hundred and thirty souls, some fifteen are Danes, and the remainder almost entirely half-breed Eskimos; not more than seven full-blooded natives are recognized among the inhabitants, of which number is the Frederick already referred to. A first impression of this singular settlement is not calculated to inspire enthusiasm for a prolonged residence in the "land of desolation." A few wooden structures, comprising a church, the government building or general store, and the residences of the Danish officials, together with a somewhat larger number of green-grown and chimneyed turf huts of the Eskimos, crown a dreary expanse of granite and syenite, over whose surfaces the ice of former ages ploughed its way to the sea. Everywhere the effects of past glaciation are plainly written. No trees of any kind shadow the sunlight from a perpetual summer sun; no song of bird, save the occasional chirp of the snow-bunting and wheatear, responds to the wakening calls of morning. The melancholy bark of a dozen or more of shapely curs—not, however, the awe-inspiring and night-destroying howl of books of travel, but the more subdued tones of reality—alone indicates possession of the town. Cheerfulness, save in the bright sunshine which here illumines all nature, seems to have forever deserted the locality.

But this first impression almost immediately disappears through closer acquaintance. Once the foot has been set upon the mirrored rocks, the charms of this garden spot one by one unfold themselves. The little patches of green are aglow with bright flowers, rich in the colors which a bounteous nature has provided; the botanical eye readily distinguished among these the mountain-pink, the dwarf rhododendron, several species of heath, the crow-foot, chickweed, and poppy, with their varying tints of green, red, white, and yellow. Gay butterflies flit through the warm sunshine, casting their shadows over "forests" of diminutive birch and willow. Here and there a stray bee hums its search for sweets among the pollen grains, while from afar, woven through the music of gurgling rills and brooks, come the melodious strains of thousands of mosquitoes, who ever cheerfully lend their aid to give voice to the landscape. Above this peaceful scene tower the dark-red cliffs of basalt, which from a height of two thousand feet look down upon a sea of Mediterranean loveliness, blue as the waters of Villafranca, and calm as the surface of an interior lake. Over its bosom float hundreds of ice-bergs, the output of the great Jakobshavn Glacier, fifty miles to the eastward, scattered like flocks of white sheep in a pasture. Such was the summer picture of the region about Disko as it was found by the writer in two successive seasons. There was little of that Greenland about it which we habitually associate with the region, nothing of those terrors which to the average mind reflect the quality of the Arctic world.

Dreary though a long residence may prove to be at a spot like Godhavn, there is yet seemingly enough comfort in it to make it attractive to the Danish officials who reside there. The neat little cottages, well supplied with those appliances and adjuncts—such as a library, piano, and billiard-room—which conduce to a home-like comfort, are not in absolute harmony with their surroundings, but they bear testimony to an intelligence and refinement governing the household which come with a rude shock to those who had expected to meet with at best only half-barbarians in this remote quarter of the globe. It was an almost inexpressible pleasure for me to see the geraniums, fuchsias, and roses which the good people were here raising behind double windows or under glass covers, and fondling with a care only equalled by the interest with which they pursued the general subject of Greenland zoölogy, or followed the recent explorations of men like Ryder, Stanley, Holub, and Peters. Herr Inspector Andersson, whose hospitality I had already enjoyed the summer previous, was absent at the time of our visit, having but a few days before gone to Upernivik to adjust some matters in connection with the government there. Mrs. Andersson and her daughter, however, gave us a kindly welcome, which was reinforced through the good offices of the Governor and his assistant. A determination to aid our expedition to the fullest extent possible was made manifest from the moment that our arrival was officially announced.

We secured some fur clothing for our equipment, and what we thought to be of greater importance to ourselves, the services of an Eskimo interpreter and servant, Daniel Johannes Matthias Isaiah Broberg, a nephew of the wealthiest native of Godhavn, and brother of Nicholas Broberg, who in 1883 acted in a like capacity for the second Greely Relief Expedition. Daniel, like most of the Eskimos of Godhavn, was inordinately fond of his tobacco, and it was rarely that he was to be found without his pipe; speaking, eating, or sleeping, his pipe appeared to be his most faithful and constant companion. The stipulations of our contract with him were, that he was to receive £3 10s. per month; that he was not to receive any orders from the ship's men; not to be obliged to draw, by himself, a sledge over the inland ice; to be remunerated for the breakage of an arm or leg, or for other bodily mutilation; to be returned to Godhavn. These stipulations, which were exacted from a fear of ill-treatment engendered through experiences associated with former expeditions, and which have made it all but impossible to secure the services of any of the Eskimos of the Inspectorate, were supplemented with a special recommendation for a pair of pantaloons.

At 1.30 P. M. of the 16th we fired our parting salute, and dipping our colors to the ship Constancia, which was then lying in port, slowly withdrew from the shadow of the tall cliffs which give to the harbor its most impressive aspect. Our destination was Upernivik, the most northerly of the Danish settlements, and the most northern settlement of civilization on the surface of the globe. We remained here but a few hours, our sole purpose being the exchange of civilities with the Danish officials resident there. Herr Inspector Andersson and Governor and Mrs. Beyer extended to us an open-hearted welcome, and with it the full hospitality which their house offered.

A more exquisite day than that which marked our departure from Upernivik could scarcely be conceived. The white lumps of ice which almost choked the harbor, and the glare from whose surfaces fairly dazzled the eye, were a marked contrast to the delicious warmth which was supplied by an Arctic $52°$.

Desolate fogs, however, broke in upon the evening and night, and it was not until two o'clock of the following afternoon (the 19th) that we were enabled to make a landing on the outer Duck Island. The Devil's Thumb, that most notable landmark, 2,347 feet in elevation, on the western coast of Greenland, should have been made before midnight; but the ice-bound fogs obliged a halt throughout the greater part of the evening and night hours. The twentieth of the month, the day that had been fixed upon for our arrival at Melville Bay, actually found us there, and we stood confronting the northern ice.

No real difficulty was encountered in the passage of this much dreaded region of the Arctic waters. An accumulation of shore-ice prevented us from following the coast in the track of the daring whalers, but about twenty-five miles seaward comparatively little heavy ice, beyond broken and rotten pans, was encountered, and were it not for a continuous lowering fog, little hindrance to a free navigation would have been presented. My note-book thus narrates our progress :

"We find solid shore ice stretching far out to sea, through which it is impossible to find a passage. Attempt a more westerly course, some twenty-five miles from shore, and enter open water. At 8 o'clock (A. M.) we are making good headway; thermometer 33°. The water is now as smooth as a mirror, with only the smallest ripples to break its surface. Icebergs glisten at almost all points of the compass, but they are not numerous. An ominous "ice-blink" shows up in the west, and under it the mate announces much ice. It seems after all that we are not destined to make an absolutely clear passage of the Melville Bay. At 10.30 we strike outliers of the ice—rotten cakes, which we cut through without mercy. The ice, however, thickens, and by eleven we get into uncomfortable quarters with a solid unbroken pan, from whose presence we ignominiously retreat. We try still further to the westward, rounding the pan, and are again in the open sea, or rather in a lead one or miles in width between the two pans. The ice, while unbroken, is not thick—one and a half to two feet—and has rotted above and below. Large bergs are moored in its midst, holding it, doubtless. Shortly before noon we are momentarily halted by the ice; start again, pushing around the cakes, and taking a generally northwest course. Too hazy for noon observations; we haul up again at about 4 P. M., and tie on to an ice pinnacle. We are nosed to a very large unbroken pan, which is rotten at its borders, but thick, eight to ten feet or more, further away. Mr. Dunphy, our eagle-eyed second mate, shouts "bear," and immediately there is a general rush for the guns. The object responding to the announcement from the crow's nest is a blackish looking mass about three quarters of a mile in our lead. Bear tracks show up on the porte, but the object of our attention gradually passes through a varied transmutation from bear to walrus, from walrus to seal and finally to an interrogation point. After tea Bryant, Mills, Entrikin and one of the ship's men put off on the ice after seal, but come back unsuccessful. We take fresh start about 8 o'clock, but do not continue long, owing to fog and ice; the thermometer has stood pretty well at 33°.

"The fog lifts at midnight and shows clear water all around. Our engines are again put into operation and the ship gets to her course at 3 A. M., continuing at full speed for about four hours, when the fog once more shuts down upon us. Scores of seals lie about us on the distant floes, but unfortunately the condition of our surroundings renders them inaccessible. We grope along slowly, hoping by chance to avoid the more formidable ice-pans which have been cast loose from their moorings and are now freely navigating in the course of the prevailing winds and currents. Every break in the fog is taken advantage of for a fresh spurt, but seemingly each effort only adds another line to the zig-zag which we have been constructing. High flat-topped bergs, the largest possibly 200-250 feet in height, loom up in the distance, and their grouping suggests that they occupy the well-known McClintock ground of stranded bergs."

At 8 A. M. of July 22d we were off Cape York, and had completed the passage of the Bay; the high land was first sighted shortly after midnight, but beyond a momentary appearance, it remained shrouded in the heavy fog until the early hours of morning. Gray cliffs of granite, moss-grown and grass-grown on their favored slopes, with here and there a glacier peacefully slumbering in their deeper hollows, mark the exit from the ice-bound Melville Bay to the open North Water. For sixty hours after leaving the Duck Islands the condition of the weather had been such that no observations for position could be taken; our course had been one solely of compass and dead-reckoning. Considering the sluggishness of the compass in these regions, and the almost endless number of detours which a course in the fog among the ice-pans necessitates, one could not but be impressed by the general directness of the traverse, and the exactitude with which it was terminated. Barely fifty hours were required for the passage from the Devil's Thumb to Cape York, and had there been no fog, even with the large quantity of ice that was present, the time would probably have been reduced by from fifteen to twenty hours.

ESKIMO CHILDREN OF SOUTH GREENLAND.

At the Eskimo settlement a few miles to the eastward of Cape York—the settlement commonly known as that of Cape York—we obtained the first information regarding the Peary party. A shaggily-bearded Eskimo, one of the tallest and most stalwart of the tribe of so-called Arctic Highlanders, measuring little (if anything) less than six feet in height, had passed some part of the winter about the "Peary igdloo" on McCormick Bay, and consequently could state something from personal knowledge. Our extremely limited acquaintance with the Eskimo tongue, combined with the difficulty with which our interpreter grasped the sense of the northern dialect, made progress in a mutual comprehension slow and wearisome; but enough was made clear that at last accounts, extending back to a period of some four or five months, the members of the party—all of whom were indicated by name—were doing well. A rude drawing, representing with fair precision the geographical contours of the region, showed that they were at that time still on McCormick Bay, and provided with both boats and sledges. Coupled with this information we were made to understand, as, indeed, we had already known previous to our departure—that one of the vessels of the Melville Bay whaling fleet had been crushed in the ice.

The arrival of the Kite at this first outpost

of the northern Eskimos was the signal for much quiet happiness on the part of the natives. Scarcely had the vessel made fast to a cake of ice before she was boarded by the happy people—men, women, and children—who, true to the instincts of an honest nature, required no invitation to bid them welcome. They stayed until they had satisfied every curiosity, or until the steam whistle announced the prospective departure of the "Oomeakshua" —the "big woman's boat," as the natives style every large vessel. Among the visitors I recognized a number of familiar faces, but the majority of my associates of last year seemed to be absent. A limping old man who had been known to Hayes was dead, and other members of the tribe had departed.

A special purpose in calling at the settlement of Cape York, or Ignamine, was the distribution among the natives of gifts of charity which had been generously contributed by citizens of Philadelphia and West Chester. Boards cut to the length of sledges, strips for kayak frames, hardware, and utensils of various kinds, cooking implements, etc., were a part of the bountiful cargo that was to give joy and wealth to a rugged people—a people to whom a barrel stave or a needle was an almost priceless treasure. Words fail to describe the scene of animation which marked the bestowal of the awards. There were no rude attempts to obtain possession of any special article, no boisterous demonstrations of superiority; each man or woman received his or her gifts with a dignity and calm composure which were truly remarkable, in view of the wealth which the presents conveyed. Their expression of extreme delight was told in a few syllables "Na, na, na, nay."

The Eskimos of the Melville Bay region and of the further north, the *Itaner* or Etahnes of Bessels, to whom Ross, with singular inappropriateness, applied the name of Arctic Highlanders—the Eskimo rarely leaving the low lands of the sea-border—are, contrary to common supposition, undergoing no diminution in numbers. Indeed, if the estimates given by earlier explorers are at all to be relied upon, they must be increasing in numbers. Hayes, in 1860, placed the total population between Cape York and Etah, the most northern of the native settlements, at about 100 souls, but Bessels, twelve years later, with the advantage of an extended period of observation, assumed that 150 would more nearly approximate the truth. The exceedingly accurate work of Dr. Cook, the ethnologist of the Peary Expedition, whose census comprises not only the number, but the names, relationships, and points of location of most of the members of this exceedingly interesting people, increases the number to 233, all of whom are at this time located between Cape York and the northern shores of Robertson Bay, or between the parallels of 75° 56' and 77° 55' (77° 45'?) north latitude. The fact that these Eskimos are of a migratory habit, changing their location as the conditions or the necessities of the chase demand—appearing now at one point, and disappearing at another—may have given rise to the notion of impending extinction, and has, doubtless, been in a measure the cause of the discrepancy in the estimates furnished by different travelers. When the Kite put in at the Cape York settlement in the summer of 1891 there were not less than sixty natives settled there, of which number forty-five were at one time on our vessel; at the time of the visit of the present expedition barely more than one half of this population remained, a northerly migration, principally to the settlement on Barden Bay (Netlik or Netchiolumi, not Ittiblu, which is some fifteen miles further to the east), having robbed the locality to the advantage of another. At the locality just named we found a population of 41 as against the twelve of the year before. On the other hand, none but "departed" traces were found at Port Foulke and on Sonntag Bay of the so-called Etah and and Sorfalik Eskimos, whose pleasant association with the expeditions of Kane, Hayes and Hall forms so bright a chapter in the history of Arctic exploration.

After a delay of a few hours, necessitated in part by the fog, the Kite pushed into the North Water, where no floes or pack-ice were encountered. The clouds hung low over the Crimson Cliffs, but beneath them scattered patches of snow, tinted at intervals with the red of the Protococcus nivalis, clung lazily to the mountain slopes. Passing Conical Rock at midnight, the expedition steamed to Wolstenholme Island, on the western spur of which it had been prearranged that records should be left by Mr. Peary, in the event of a forced early retreat, but no cairn was discovered. My own advice of the prospective Relief Expedition, which had been deposited on the same island nearly six weeks earlier (June 13th), by Captain Phillips, of the whaler Esquimaux, was picked up by my men and found to be undisturbed. The party of exploration had manifestly not yet passed to the south. Shortly after 5 A. M. (of the 23d), the Kite shaped her course to Whale Sound, and early in the evening of the same day, after discharging a second cargo of charities to the Eskimos of Barden Bay, made the passage between Northumberland and Herbert Islands. Throughout the greater part of the day there prevailed a balmy and spring-like temperature which was in striking harmony with the warm, sunlit effects which the landscape everywhere presented. We were less than nine hundred miles from the

Pole, yet the thermometer could not be coaxed down even to the freezing-point; in the sun the mercury rose rapidly to near the 60° line. Thousands of ice fragments, thrown out by one of the arms of the great Tyndall Glacier, covered the silvered surface of the sea; while off in the distance swung out in majestic line the flotilla of bergs to which the giant glaciers of Inglefield Gulf have given birth. Murchison Sound was reached at ten o'clock, and only ten miles now intervened between our ship and the spot where, a year before, the "West Greenland" party saw fashioned the wooden shelter which was to give lodgement to the brave seven who composed the Peary party. Expectancy is now at full height, and from every point of vantage on the vessel comes the desire to possess the eyes that see the first and farthest. The bow, the rigging, the bridge, and crow's nest are all in active competition, but the award of victory is to be withheld for some time as yet. McCormick Bay opens up broadly to the east, its moving ice-field joining with the endless fleet of bergs which are slowly coursing to the open sea. Five miles more are covered, and the Kite plunges into the soft pack, but no sign of human life or habitation is as yet apparent. Through the clear night air is sent the boom of the ship's cannon, but only reverberations from the barren crags answer. Save the occasional crackling of a feeble iceberg, and the noise of the ship's machinery, all is as quiet as the grave. A second discharge follows, accompanied by the shrill tones of the steam-whistle, but still no answer. The red cliffs of Cape Cleveland are now near to us, and the range of vision, except for an intercepting berg, covers the site which we know to be that of the Peary igdloo. Presently from far aloft comes the welcome: "They are answering us with a gun." No sound was audible, but the keen eye of Second Mate Dunphy had detected smoke. Three long shrieks from our siren, as a token of welcome, and the pennant swings to the breeze. When the ship's thunder once more broke the ominous silence a small speck appeared upon the water's surface. "They are coming to meet us in a boat," came the cry from aloft, and the field-glass confirmed the observation from the crow's nest. In the nearing boat were Verhoeff, Cook, and Gibson, who had come with Eskimo friends to greet the strange apparitions from the south. A half hour before the midnight hour they boarded our vessel, and we obtained from them the happy tidings that everything was well. Lieutenant Peary, who had entirely recovered from the accident of last summer, was at the time of the arrival of the Kite, with young Astrup, traversing the vast wilderness of the inland ice, while the heroic wife of the commander, with Matthew Henson, was encamped at the head of the bay, some fifteen miles distant, awaiting the return of the explorers.

The members of the Peary party who had come out to meet us showed no signs of a struggle with a hard winter. Their bronzed faces spoke more for a perpetual tropical sunlight than for a sunless Arctic night, the memories of which had long since vanished as a factor in their present existence. No serious illness of any kind had invaded the household during a twelve months absence from civilization. The expedition quarters presented a very different appearance from what they did a year before when the Kite steamed out from McCormick Bay. The diminutive two-roomed house, which then stood solitary and uninviting in its own field of scattered mountain-pink and poppy, roofless to the elements and unprotected from the blasts which were hurled against the sides of board and tar-paper, was now the focus of a busy world that had congregated about. A colony of Eskimos, whose members had been gathered in from various settlements along the coast, had established themselves on the same free soil of nature, eager to reap the benefits which a contact with civilization might bring, and ever ready to give a helping hand to those whom they now recognized as superiors. The twenty or more natives were lodged in five tupics, or skin summer tents, about which were gathered a variety of paraphernalia necessary to the Eskimo household and an amount of odor which only weeks—more likely months—of abrasion and ablution could efface. If cleanliness was not a virture with these people, their honesty, cheerfulness, and good-will made amends for the lack of a quality which a defective vision has assigned to be the first attribute of Godliness. The majority of the men and women were of low stature, the tallest of the latter, fat Itushakshui, the mother of an exceedingly winsome young bride of thirteen, Tongwingwa, measuring only 4 feet 8 inches. M'gipsu, the shortest of the mothers, measured only 4 feet 4 inches. The men are, with few exceptions, taller than the women, but even among them a stature exceeding five feet is a rarity rather than the reverse, although such exceptional cases are less rare among the people of the region about Cape York than further northward.

The moment that the Kite appeared in McCormick Bay the natives recognized that a "circus had come to town." A few of them had seen the vessel, or one similar to it, before, but to the majority the Oomeakhshua was an unimaginable novelty. At all hours of "night" and day, when a transfer could readily be made from the shore, men, women, and children would gather to her sides, eager to obtain mementos of our journey in the shape of biscuits, soup, or thimbles. The deck and cabins underwent a daily inspection, as did also the forecastle and every other available spot of interest which the ship offered. These visits to us ultimately became a source of some annoyance, since they interfered largely with the work—the making of skin boots and clothing, fashioning of sledges and kayaks, etc.—which had been laid out for them by the Peary party. So long as the vessel was in sight and approachable, it formed the uppermost thought in their minds, more especially of the women. Stitching seal-boots, or kamiks, or chewing hides to render them pliable, was of little moment so long as good-hearted Captain Pike gave them welcome with him, and dealt out rations of bread and biscuit. On two occasions we were favored with a song and dance, the instrumental accompaniment being given on a stretched drum-like hide, the frame of which was beat to a three-time with a splinter of ivory. The most popular melody—the one which is supposed to have curative powers when sung by the "angekoks" or wise men of the settlement—consisted of a succession of yah, yah, yahs, and scarcely anything more, which fell in rhythmic cadence from a high crescendo to a tremulous under-note, suggestive of almost any range of possibilities.

Almost immediately after our arrival a message was sent up by special Eskimo express to

Mrs. Peary, informing her of our coming, and in a few short hours a welcome greeting was returned to the relief party. I visited her camp on the following day (25th). The bay was still largely closed with ice, and the upper part was accessible only by way of the long shore line, on which a lingering ice-foot had set its heavy masses of frozen sea. Just outside the tent, in the midst of a mosquito-tract which, for the quality and quantity of its musical tenants, could readily vie with the more favored spots of the tropics, I met the brave woman who was the first of her race to dare the terrors of the north Arctic winter. She had come to meet me and pressed a cordial invitation to follow to her cosey shelter. The little white tent, whose only furniture consisted of two sleeping-bags of reindeer-fur, stood on a patch of meadow-land facing the bay and across it the bold granite bluffs which to the outer world marked the last traces of the departed explorers, and over whose nearly vertical walls it was hoped that fortune would favor an early return. A range of steep heights, over whose declivities a number of glaciers protruded their arms caterpillar-like in the direction of the sea, formed the desolate background. Eastward the eye gazed upon the interminable ice-cap, with its long sweep of gentle swells and undulations—a land lost between the sky and the earth; westward it fell upon the broad expause of the bay whose half-congealed surface passed hazily to the distant sea beyond. This was the picture of the spot where Mrs. Peary, almost alone among the few wild flowers by which she was surrounded, had passed full nine days with but a single companion to help relieve the dreary and anxious hours of waiting. The experiences of a year had told lightly on her, and there was nothing to indicate regret for a venture which no woman had heretofore braved and which only noble devotion had dictated.

Recognizing, with the late day of his departure from McCormick Bay (May 1st), that Mr. Peary could not readily return from his hazardous journey before the first week of August, and that no purpose would be subserved by the relief party remaining at their present quarters until that time, I ordered out the Kite on the following morning to proceed to Smith Sound, hoping that a fortunate combination of circumstances might permit us to make a traverse of the front of the great Humboldt Glacier. In this hope, however, we were destined to be disappointed. No more delightful weather could have been conceived than that which marked the day of our departure northward. A flood of light poured over the landscape, illumining it with a radiance which only the snows and ice of the far north or of Alpine summits can reflect. Scarcely a breath of air disturbed the hundreds of bergs and "berglets" which floated lazily by, impelled by the gentle current of the deep blue sea, and barely a ripple, save where the little auk had congregated in hundreds to disport awhile in the warm sunshine, broke the surface of the mirror into whose inner depths we cast our images. Fifty miles northward the headland of Cape Alexander stood out with a boldness that was almost startling in its effects, while beyond it a few minor heights marked the passage into that forbidding tract of sea and ice from which so many brave hearts have never returned. Before we had reached Littleton Island the ominous ice-blink only too plainly told us that ice was ahead; Smith Sound was closed from Greenland to the American side. At midnight we were brought up by the "pack"; Cape Sabine, memorable in the annals of Arctic discovery as the scene of disaster and of heroic rescue, was to our left, and Rensselaer Harbor, equally memorable as the winter quarters of the Advance of Kane, a few miles to the eastward. The ice was somewhat heavier than the "pack" of Melville Bay, in which we were imprisoned the summer previous, but it yet bore the same quiet and tranquil air, wholly unsuggestive of power-possession. The hummocky sheets, measuring from six to ten feet in thickness, and showing but a single lead in their midst, had manifestly not yet begun to break for the season, and therefore all efforts to reach the glacier at this time must be fruitless. Although nine years had elapsed since the crushing of the Proteus, the experiences of that desolate July 23d were still too vivid in the mind of our captain to permit of any risks being taken on this occasion. With his back turned to the snow-clad slopes of Cape Sabine, and gazing upon the uncovered and to him less reminiscent heights of. the Greenland coast, he announced that we had reached the journey's end. The Humboldt Glacier was invisible, although farther off to the northward, the prominences of Grinnell Land, Capes Hawkes and Louis Napoleon, and possibly also that of Cape Imperial, carried the eye quite to the border line of, or even beyond, the eightieth parelel.

The front ice of the Smith Sound pack is the home of the walrus. Hundreds of these animals were disporting themselves in the silent hours of a sunlit midnight; here a few gathered on tablets of floating ice, others leisurely paddling about with an abandon truly majestic. Their frolics immediately called to mind the gambols of pups and kittens. No animal, probably, save the Bengal tiger, offers the same amount of sport to the huntsman as does this king of the northern waters. Every attack resulting in a wounded animal can be safely relied upon for a counter-attack, which is prosecuted with an audacity no less remarkable than the energy with which it is sustained. A wounded walrus will not infrequently call for assistance to a number of its associates, and woe be then to the huntsman if, in the general struggle, one of the infuriated animals should place its tusks on the inner side of the little craft that has gone out to do battle.

The largest specimen secured by us measured, from the tip of the nose to the extended hind flippers, somewhat more than thirteen feet (to the extremity of the spinal column, eleven feet four inches); its weight was estimated to be between fifteen hundred and two thousand pounds, but not impossibly it was considerably more.

In our return southward to McCormick Bay, which began shortly before five o'clock of the morning of July 27th, explorations were extended into Port Foulke and Sonntag Bay, where were located the "tribes" of the Etah and Sorfalik Eskimos, the most northerly of the inhabitants of the globe. Only empty huts, five or six at each locality, a few graveheaps, and distributed rubbish of one kind or another, now indicated a former possession of the land; adverse conditions of the chase had driven away the inhabitants, who had departed south to add their little mite to the colonists of the Whale Sound region. The last of the Etahs had joined the cantonment about the Peary igdloo. That the region of Port Foulke had only recently been abandoned was proved by the generally good state of preservation of the stone huts, not less than by the newly arranged fox-traps that were outlying. A return of the departed could probably be expected in a more propitious year. In Sonntag Bay an effort was made to ascertain the possibilities of some of the large glaciers as a means of communication with the upper ice or ice-cap. The fact that in many of these northern ice-streams crevasses were largely or almost entirely wanting, or were so completely closed as to show but mere rifts on the surface, seemed to indicate that a direct highway of travel, accessible alike to sledge and man, could be found on the moving ice. A first attempt on a northeast glacier, with a sledge loaded to about two hundred pounds, proved abortive; the high terminal wall and abrupt lateral slopes, while they offer no serious hindrance to man in the capacity of pedestrian, blocked the approach of the toboggan, as would, indeed, have also done the numerous crevasses which cut across the ice in its lower border. A second attempt, made on the glacier* discharging into the eastern extremity of the bay, proved more success-

---

*It is with special pleasure that I associate with this undescribed glacier the name of George W. Childs, Esq., of Philadelphia, to whose kindly interest in the expedition under my command the expedition owes much for its organization and success.

ful. Ascending over the feebly depressed lateral moraine of the left side, no difficulty was encountered in transferring our impedimenta to the surface of the glacier, which was practically solid, and almost without rift for miles from its termination. The even crust of the ice, which at the early hour of twelve had barely begun to yield to the softening influences of a midnight sun, offered little obstacle to the traction of our sledge, and before five hours had passed, we had planted our stakes in the névé basin, 2,050 feet above the sea. A portion of the immediate ice-cap was below us, some of it eighty or a hundred feet higher up; the feasibility of the passage had been demonstrated.

Later experiences on some of the more southerly and still more gigantic glaciers only further demonstrated the accessibility of the ice-cap along a route of travel where the gradient was scarcely ten degrees, and in many parts considerably less. Indeed, the slope of many of the northern glaciers for miles does not exceed three to five degrees. The only difficulty that we encountered in the traction of our sledge, a steel runnered toboggan, over the Sonntag Bay glacier was on the upper course of that ice-sheet. The hard and slippery crust which there rises into a series of rapidly flowing swells or mounds, in some places disposed in almost amphitheatric regularity, was a hard test to the almost unshod condition of our walking gear. We had unfortunately provided ourselves with only indifferent creepers or climbing irons, whose penetration in the ice was not sufficient, with the dragging weight behind, to secure lodgement or resistance to the feet. It required the full force of the six men of our party to mount these ice-hillocks, the accomplishment of which would otherwise have been a comparatively easy matter. On the down-journey, again, the tendency of the toboggan to run by itself necessitated a continuous "brake" to be applied to it in the form of guiding lines, which were held on either side by the members of the party. In the upper course of the glacier, where the slope mounts more abruptly to the forming basin, a number of crevasses were met with, but with the exception of a few which had to be circumvented, they were nearly all easily crossed, the toboggan in some cases being pushed over to form a bridge, over which the least gymnastically-inclined of the party found a ready passage. We found it, however, a wise precaution to use the Alpine rope, and indeed, had the satisfaction of seeing this precaution emphasized by a plunge which Mr. Bryant took into one of the crevasses. Fortunately a deep descent was prevented by the narrowness of the fissure and by an accumulation of ice-debris which clogged and disfigured

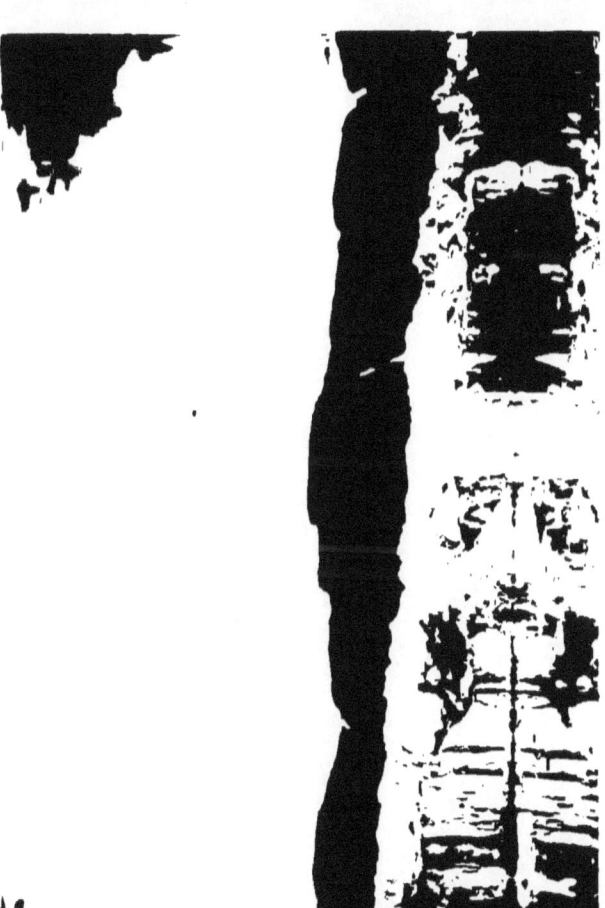

TERMINAL WALL OF VERNOIRE GLACIER.

the interior, almost closing it to the top.

We arrived at our old quarters in McCormick Bay in the evening of the 29th. The balmy weather that had thus far accompanied us still gave the sensation of spring, but an impending change was perceptible. The last two or three evenings had grown measurably cooler, and the drooping sun indicated a drawing approach to cold weather and wintry nights. Anticipating a probable return of Mr. Peary toward the close of the first week in August, the Kite, with Mrs. Peary and Matthew Henson added to my party, steamed on the 4th to the head of the bay and there dropped anchor. Mrs. Peary accompanied me in the afternoon to the foot of one of the glaciers which here debouches in the valley, terminating about a quarter of a mile from the water. It was my intention to measure the velocity of this ice-stream, and toward that purpose a number of sight stakes were planted on the off-border; unfortunately, circumstances were such as not to permit a return to the same locality, and, therefore, no record was obtained. Like seemingly all the glaciers, at least those of a second order, of north Greenland, this was one of recession; the open space between it and the bay was everywhere marked by evidences of former glaciation, with here and there traces of terminal moraines still standing. Naturally, it could not be ascertained whether actual recession was still in progress, or whether a reversed condition had possibly set in; only the observations of a succession of two or more years could determine this point.* Vast quantities of water were being everywhere thrown off from the ice, some of it falling in great lateral fountains which cut ragged gashes into the solid wall of ice. The heat of a single summer, being continued almost equally through night and day, graves deeply into these northern ice-sheets, and the effects of lateral ablation are especially well marked. Many of the glaciers are bordered by lateral ravines, portions of the glacial trough which have been left exposed by the disappearance from them of the excavating ice. The surface waste is surprisingly great in some instances. One of the arms of the great Tyndall Glacier which on our journey northward was found to encompass the north side of Bell (Fitzclarence) Rock, in Booth Sound, had all but disappeared in the interval of five weeks preceding our return.

On the day following our arrival at the head of the bay, a reconnaissance of the inland ice,

---

*The examination of the much larger glacier, since named by Mr. Peary the Sun Glacier, which discharges, with a frontage of a mile, on the opposite side of the bay, gave an advance of the lateral margin of nearly seven inches in eight hours; on the other hand I could detect no movement, for a period of twenty-four hours, in the large glacier which descends from the Cleveland plateau toward Murchison Sound.

with a view to locating signal posts to the returning explorers, was made by the members of the expedition. A tedious half-hour's march over boggy and bouldery talus brought us to the base of the cliffs, at an elevation of three hundred and fifty to four hundred feet, where the true ascent was to begin. The line of march is up a precipitous water-channel, everywhere encompassed by boulders, on which, despite its steepness, progress is rapid. The virtual crest is reached about six hundred and fifty feet higher, and then the gradual uprise of the stream-valley begins. Endless rocks, rounded and angular—the accumulation of former ground and lateral moraines—spread out as a vast wilderness, rising to the ice-cap in superimposed benches or terraces. At an elevation slightly exceeding eighteen hundred feet we reached the first tongue of the ice. Rounding a few outlying "nunataks"—uncovered hills of rock and boulders—we bear east of northeast, heading as nearly as possible in the direction from which, so far as the lay of the land would permit us to determine, the return would most likely be made. The ice-cap swells up higher and higher in gentle rolls ahead of us, and with every advance to a colder zone it would seem that the walking, or rather wading, becomes more and more difficult. One by one we plunge through the yielding mass, gasping for breath, and frequently only with difficulty extricating ourselves. The hard crust of winter had completely disappeared, and not even the comparatively cool sun of midnight was sufficient to bring about a degree of compactness adequate to sustain the weight of the human body. At times almost every step buried the members of the party up to the knee or waist, and occasionally even a plunge to the armpits was indulged in by the less fortunate, to whom perhaps a superfluity of avoirdupois was now for the first time brought home as a lesson of regret. We have attained an elevation of 2,200 feet; at 4 P. M. the barometer registers 2,800 feet. The landscape of McCormick Bay has faded entirely out of sight; ahead of us is the grand and melancholy snow waste of the interior of Greenland. No grander representation of nature's quiet mood could be had than this picture of the endless sea of ice —a picture of lonely desolation not matched in any other part of the earth's surface. A series of gentle rises carries the eye far into the interior, until in the dim distance, possibly three-quarters of a mile or a full mile above sea-level, it no longer distinguishes between the chalky sky and the gray-white mantle which locks in with it. No lofty mountain-peak rises out of the general surface, and but few deep valleys or gorges bight into it; but roll follows roll in gentle sequence, and in such a way as to annihilate all conceptions of space and distance. This is the aspect of the great "ice-blink." It is not the picture of a wild and tempestuous nature, forbidding in all its details, but of a peaceful and long-continued slumber.

At 5.45 P. M., when we took a first luncheon, the thermometer registered 42° F.; the atmosphere was quiet and clear as a bell, although below us, westward to the islands guarding the entrance to Murchison Sound, and eastward to a blue corner of Inglefield Gulf, the landscape was deeply veiled in mist. Shortly after nine o'clock we had reached an elevation of 3,300 feet, and there, at a distance of about eight miles from the border of the ice-cap, we planted our first staff—a lash of two poles, rising about twelve feet and surmounted by cross-pieces and a red handkerchief. One of the cross-pieces read as follows: "To head of McCormick Bay—Kite in port—August 5, 1892."

A position for a second staff was selected on an ice-dome about two and a half miles from the present one, probably a few hundred feet higher, and commanding a seemingly uninterrupted view to all points of the compass. Solicitous over the condition of the feet of some of my associates, I ordered a division of the party, with a view of sparing unnecessary fatigue and the discomfort which further precipitation into soft snow entailed. Mr. Bryant, in command of an advanced section, was entrusted with the placing of the second staff, while the remaining members of the party were to effect a slow retreat, and await on dry ground the return of the entire expedition. Scarcely had the separation been arranged before a shout burst upon the approaching midnight hour which made everybody's heart throb to its fullest. Far off to the northeastward, over precisely the spot that had been selected for the placing of the second staff, Entrikin's clear vision had detected a black speck that was foreign to the Greenland ice. There was no need to conjecture what it meant: "It is a man; it is moving," broke out almost simultaneously from several lips, and it was immediately realized that the explorers of whom we were in quest were returning victoriously homeward. An instant later a second speck joined the first, and then a long black object, easily resolved by my field-glass into a sledge with dogs in harness, completed the strange vision of life upon the Greenland ice. Cheers and hurrahs followed in rapid succession—the first that had ever been given in a solitude whose silence, before that memorable summer, had never been broken by the voice of man.

The distance was as yet too great for the sound to be conveyed to the approaching wanderers, but the relief party had already been detected, and their friends hastened to extend to them a hearty welcome. Like a veritable giant, clad in a suit of deer and dog-skin, and gracefully poised on Canadian snow-shoes, the conqueror from the far north plunged down the mountain-slope. Behind him followed his faithful companion, young Astrup, barely more than a lad, yet a tower of strength and endurance; he was true to the traditions of his race and of his earlier conquests in the use of the Norwegian snow-skate or "ski." With him were the five surviving Eskimo dogs, seemingly as healthy and powerful as on the day of their departure.

In less than an hour after Lieutenant Peary was first sighted, and still before the passage of the midnight hour of that memorable August 5th, culminated that incident on the inland ice which was the event of a lifetime. Words cannot describe the sensations of the moment which bore the joy of the first salutation. Mr. Peary extended a warm welcome to each member of my party, and received in return hearty congratulations upon the successful termination of his journey. Neither of the travellers looked the worse for their three months' toil in the interior, and both, with characteristic modesty, disclaimed having overcome more than ordinary hardships. Fatigue seemed to be entirely out of the question, and both Mr. Peary and Mr. Astrup bore the appearance of being as fresh and vigorous as though they had but just entered upon their great journey.

After a brief recital of personal experiences, and the interchange of American and Greenland news, the members of the combined expedition turned seaward, and thus terminated a most dramatic incident. A more direct meeting than this one on the bleak wilderness of Greenland's ice-cap could not have been had, even with all the possibilities of prearrangement.

At 4.30 of the morning of August 6th Mr. Peary met his devoted and courageous wife; and on the following day, in the wake of a storm which grounded the good rescue ship and for a time threatened more serious complications, the Kite triumphantly steamed down to the Peary winter-quarters at the Redcliffe House.

The results of the Peary Expedition justify all the anticipations that had been pinned to it. Apart from its worth in determining the insularity of Greenland—thereby setting at rest a question which had disturbed the minds of geographers and statesmen for a period of three centuries, or since the days of Lord Burleigh— it has forever removed that tract from a consideration of complicity in the main workings of the Great Ice Age. The inland ice-cap, which by many has been looked upon as the lingering ice of the Glacial Period, stretching far in-

THE VERHOEFF GLACIER.

to the realm of the Pole itself, has been found to terminate throughout its entire extent at approximately the eighty-second parallel; beyond this line follows a region of past glaciation—uncovered to-day, and supporting an abundance of plant and animal life not different from that of the more favored regions southward. Over this tract has manifestly been effected that migration of organic forms from the west and to the west which has assimilated the faunas and floras of eastern Greenland with those of other regions; indeed, man's own migrations are probably bound up with this northern tract. Significant, too, is the discovery of giant glaciers passing northward from the inland ice-cap, and discharging their icebergs into the frozen sea beyond. The largest of these, named the Academy Glacier, and measuring from fifteen to twenty miles in width, empties on the northeast coast into Independence Bay, under the eighty-second parallel.

V.

A Lost Companion.

Shortly after the return from the interior of the exploring party, and pending preparations for the final departure southward, happened that one incident to the expedition which in any way marred the brilliancy of its exploits. It was at this time that Mr. Verhoeff, the meteorologist and mineralogist of the North Greenland party, undertook that last search after rock-specimens from which he never again returned to meet his associates. He was last seen on the morning of August 11th, when he stated his intention of visiting the Eskimo settlement of Kukan, across the northern wall of McCormick Bay, and a mineral locality well known to him. Failing to appear at an early day, fears were entertained for his safety, and a systematic and scattered search was immediately instituted by our combined parties, assisted by nine specially selected Eskimos and several members of the ship's crew. The search was extended almost unremittingly throughout seven days and nights, over mountain, ice, and glacier, and with a thoroughness that left no large area of accessible country uncovered.

For the first four days the members of my party, assisted in part by Dr. Cook, by Mate Murphy, Engineer McKinley and sailor "Jim" of the ship's crew, and four of Mr. Peary's Eskimos (Equaw, Anauka, Kooko, and Tongwe), explored the shores of McCormick and Robertson Bays, while Mr. Peary and his party made the traverse of the divide which separates these two bodies of water from one another. By this division we hoped to cover at the same time both the lowland and upland regions, and thereby lessen the chances of possible failure. Our endeavors were, however, fruitless. Neither cliff nor valley gave out any vestiges of the missing man—no trace that human feet had recently trod the frozen soil. Failing in our examination of the open land-surface, we next directed our attention to the huge ice-sheets which like so many rivers break the continuity of the shore-line, and sail out their ever-crumbling masses into the sea beyond. The possibility that Mr. Verhoeff had attempted a tour of McCormick Bay to gain the Redcliffe House, to accomplish which a traverse of the large glacier emptying at the northeast angle of the bay would have been necessitated, laid our course in the direction of this ice stream. It was early in the evening of the 19th of August, when the elevation of the sun still marked about twenty degrees above the horizon, that we again entered the shadows of the same granite cliffs over which, only a few days before, we had so joyfully passed after our meeting with Mr. Peary. The scene had changed. The deep cañon, along which the eye could follow the long lazy line of glacier for a distance of 12-15 miles to its mother ice-cap, looked bleak and forbidding; there was no longer that charm of the unknown about it which attracts when all nature smiles with success. A dark cloud had settled over the landscape, and for a time closed out its joys.

We approached the front wall of the glacier with caution and in almost silence, fearing lest any percussion might too hastily precipitate some of the tottering masses which were "calving" their way to sea as bergs. Like the snowy avalanches of the Alps, which are at times called to life by the clapping of the hands, so must these ice-masses of the north be left to their own peaceful slumbers. Once overturned, there can be no forecasting of the commotion that might follow. A turn or two may end the scene, or it can be that it has hardly begun before the water is churned into foam. We secured a safe landing in a small bight on the eastern flank of the glacier, and there hauled up our boat.

Cutting our steps into the dome-shaped lateral margin of the glacier we soon gained the surface, upon which walking was fairly easy and comfortable. An effort to reach the opposite side was frustrated by the numerous long and deep crevasses which cut into the median portion of the ice; we were obliged to wander around and about some of these, but generally could manage to keep on a united body, or where the fissures were of but insignificant width. For some distance the surface kept disagreeably hummocky, but after passing a feeding glacier, it spread out in an almost horizontal glistening sheet, admirably adapted for sledging purposes, and of necessity, to pedestrianism. The crevasses became less and less numerous, and ultimately ceased altogether, so that a traverse could be made in any direction. A narrow, remarkably straight and evenly-defined medial moraine, more in the nature of a dirt-band with angular blocks scattered over it—the like of which is to be found only in the old-fashioned books of geology, the illustrations in which are ordinarily considered to be archaic rather than truthful—occupied the central axis, stretching-off upwards to the limits of vision.

Had our mission been different from what it really was we might have said that this glacial travelling was truly delightful. With all the beauty of the ice-fields of Switzerland, and that charm of pedestrianism which an unexpected and varying change of scene carries with it, we had here the advantage of the many hours, the consciousness that a journey was not limited to any arbitrary separation of day from night. It was all day, albeit the sun shone for only a paltry few hours. For some time angry-looking clouds had been gathering about the blackened granite crests; the side cañons poured out their fleecy hosts, and before long the wild spirit of the mountains swept demon-like across the valley of the glacier. The few lazily-falling flakes which for a half-hour or so had portended evil, were before long replaced by blinding sheets of snow, and for a short time, save in its elements, nature ceased to exist. The landscape was completely blotted out from view. We were not prepared for this change, and the cold wind stung mercilessly wherever it caught an exposed surface. We muffled ourselves as best we could in our not over-generous garments, but yet it was not all solid comfort. Fortunately, the storm was of only short duration, and in its wake the landscape rose resplendent in its new garb.

At about 3 A. M. we started on the return; we had penetrated up-stream about five or six miles, and had ascended probably 6-700 feet in that distance. We had seen nothing, and no sound, save the echoes from the beetling cliffs of granite and trap which here rose in impending masses 2,500 or 3,000 feet above us, responded to the oft-repeated shouts to which we gave utterance.

On the day following our return to the Kite

CROSSING THE VERHOEFF GLACIER.

a second search was made over the same glacier, starting from the opposite side, but with no better result. Equally unsuccessful were the members of my party who had searched in other directions, and before noon Mr. Peary, who had returned from his own arduous search, only communicated further intelligence of failure. Some of his men were, however, still searching over the mountain heights towards Robertson Bay, and a ray of hope remained that they might have met with better success. With a view of combining our forces in a final effort, and of affording relief to the mountain party, the Kite was for the second time directed into Robertson Bay, the shores of which were closely scanned, but without result. Aged Kauna, the lord of the one-family settlement of Igludahominy—the last of the Etahs—whose lonely wigwam guards entrance to a patch of green which nature had specially nursed in her bosom, had neither seen nor heard of the missing man; on the opposite shore, the igdloos of Kukan were found to be all deserted and untouched. At 2 A. M. of the 22d we reached the head of the bay, and were there joined by Mr. Peary and his Eskimos; no word had yet been received from the mountain party, Astrup, Gibson, Cook and their native contingent, but it was known that their descent would be in the direction of the glaciers which break through the pinnacled walls of granite which bound the eastern extremity of the bay.

The strain of almost continuous travel during the last few days, combined with little sleep, forced a temporary rest upon us, and it was not until the hours of morning had been well advanced that we attempted to face the ice-sheet which promised only a forlorn hope. Over an extent of two full miles the deeply hollowed-out or concave ice-front presents an unbroken wall of 70-100 feet elevation, from which icebergs constantly press forward. A mile or more in advance of it the polished surfaces of three protruding rock masses or islets clearly betray a former greatness and subsequent shrinkage, but such as it is, the glacier would not be shamed by the largest of the Alps or of Scandinavia. With the arrangement that the examination of this huge ice-sheet shoud be made simultaneously from opposite sides, we divided forces, my party taking up the start from the western border. Our landing place was a true garden spot. The luxuriant growth of grass, 12-16 inches in height, with its garniture of poppies, chickweed, potentillas and gentians, was the most refreshing exposition of Greenland vegetation that we had yet met with; butterflies flitted about in the bright sunshine, whose genial warmth recalled memories of a distant south.

The back of the glacier was easily reached

by mounting over the low rounded flank which partially overrides the lateral moraine. Near to the margin the sea of ice was smooth and gently undulating, but towards the center it becomes rapidly rugged and fissured. The whole is piled up in a wild confused mass, peaks and pinnacles rising up innumerable. I ordered on the creepers and Alpine rope, and leading, was followed in the order of Meehan, Hite, Bryant, Mills, Vorse, Entrikin, and Daniel; Mr. Stokes had been disabled in the beginning of the journey by stumbling over one of the numerous holes which encumber the grass slope of the mountain-side, and had to be left in the rear. It was manifest that no crossing could be attempted in the lower course of the glacier; the disrupted surface bore a most wicked aspect, and we dared not trust to the snow-bridges which crossed the border crevasses. Into many of these, down to a depth of fifty or a hundred feet, we peered, with the thought of possibly seeing traces of our unfortunate companion, but the eye brought back with it only visions of the beautiful blue ice-walls, and of the long icy pendants which clung close to their surfaces. Striking diagonally up-stream we soon broke into a less torn portion of the ice-sheet, where travelling was comparatively easy, and where but a few crevasses disturbed the line of march. Shortly after 4 P. M., when we had just about reached the middle of the ice-stream, we met Mr. Peary and two of his Eskimos, Kumenapik and Mektosha, crossing from the opposite side. They brought to us the intelligence that the first traces of the missing man had finally been found; partially obliterated foot-prints, a few rock fragments placed on a boulder, and bits of paper from a meat-tin label, were discovered on the lateral ice which for some distance accompanies the glacier on its eastern flank. The discovery was made by the mountain party, who had so far as possible followed in the course which it was assumed must necessarily have been taken in an effort to cross the ridge. It was evident that Verhoeff had descended the cliffs and attempted to cross the glacier possibly with a view of reaching Igludahominy. The Eskimos declared the foot-prints old, and probably a full week had elapsed since they were planted on the soft snow.

At Mr. Peary's request my party proceeded up the glacier to a prominent nunatak which, about two miles further, splits the glacier into two unequal arms, one of which had already been searched by Gibson and Astrup. The glacier here expands to a width of about four miles, and presents a charm of scenery which, of its kind, I had never before seen equalled. The ice-cap is the bounding vision of the dim distance, but on either side the beetling cliffs of granite narrow the horizon to a sharp reality. We encountered little difficulty in making a median traverse of the glacier; the surface of the ice was crystalline-granular through hard freezing and alternate melting, and there were no crevasses of any account. A wilderness of small hummocks supplied their place. Occasional water-courses had graven winding channels in the ice, but they were of insignificant depth, and offered no obstacle to their passage. Shortly before seven o'clock we reached the base of the central nunatak, a giant granite mass, stained red with its rusty coat of lichen, and rising probably not less than 800 or 1,000 feet out of the ice. Its foot is buried in a garden of grass, moss and wild flowers—a veritable oasis in an ice-wilderness.

We had now reached a point on the glacier beyond which it would have been superfluous to make further search. Regretfully, therefore, we turned in the direction of the sea, with but little hope left to lighten our footsteps. We followed for some distance a medial depression in the ice (formed through the junction with, and deflection by, a secondary glacial arm—a negative medial moraine), and then deflected the course to the side of the glacier opposite to that from which we started Good areas of travel alternate with bad ones but it is easily noticeable that the crevasses rapidly increase both in number and size. Long detours are necessitated, and step-cutting be comes more frequent than pleasant. For some distance before the edge of the ice is reached the surface is frightfully torn into yawning chasms and jagged pinnacles, both alike forbidding and impassable. The snow bridges are no longer to be trusted, and over the wider chasms they are entirely absent. At 10.30 P. M we met the remaining members of the Peary party, and with them rowed out to the Kite, which we reached shortly before the midnight hour.

The aspect of the almost effaced foot-prints convinced me that the Eskimos were correct in their belief that they must have been made many days before their discovery. It was in vain that any effort was made to follow them up—there was no outlet. The search was continued through all the opening valleys and gorges of the region, and back again to the final glacier, but only with negative result Painfully we were forced to the conclusion that the unfortunate man had met his fate in attempting the passage of that wicked portion of the ice-sheet opposite which the scanty, but positive, traces of his presence had been detected.* Under this conviction, and recogniz-

---

* It is but proper to state here that a sister and uncle of Mr. Verhoeff believe the missing man to be still alive, and that he designedly separated himself from the expedition through a fondness for the life that he had been leading, and for th purpose of making a "record." No one wishes more heartily that this may be the fact than the writer of this narrative.

ing the futility of further search, the expedition returned to McCormick Bay, on the northwestern promontory of which (known as Cape Robertson), on Cairn Point, a cache of provisions was left by Lieutenant Peary.

The final departure from McCormick Bay took place on the day following the return from the search (the 24th). At 2.20 P. M. a parting salute was blown, and the Oomeakshua, whose presence had given so much joy to the rude children of the north, turned her nose homeward. Taking advantage of the fine weather, we called at the Eskimo settlement on the southern shore of Saunders Island, but found it deserted; the inhabitants were probably at their more favored retreat on North Star Bay, access to which was impossible at the time of our visit. Much ice, as a result of continuous south and southwest winds, had driven into the North Water and choked the shore passage of Melville Bay, but groping out in the direction of the "middle sea," we found our exit, and, early in the morning of the 30th, reached the first outpost of civilization, Godhavn. Without special incident, beyond the official courtesies which the expedition received at the capitals of the two Inspectorates of Greenland, Godhavn and Godthaab, and which must forever remain among our pleasurable reminiscences, the voyage was continued to the port of destination of the Kite, St. John's and thence to Philadelphia. The debarkation at the latter port was made between ten and eleven o'clock on the morning of September 23d. The mission of the Relief Expedition had been accomplished.

VI

THE GREENLAND ICE-CAP AND ITS GLACIERS.

The distinctive feature of all Greenland is its white mantle. Seen from almost any point off the coast, except where fogs and mist have unnaturally limited the horizon, the object that most distantly appeals to the eye is some portion of the interminable ice-cap, which rises above all other features of the country, and gives to it its general surface. Travellers have measured on this surface elevations of 8,000 and 9,000 feet, and not unlikely still greater elevations will yet be recorded. Rising in a measure dome-like to the interior, the "height of land" attains its full elevation only after miles of country have been traversed, and *possibly* the highest point lies not very far from the axial centre of the land. This is, however, doubtful, and there seem to be no very good grounds for believing it to be the case. Nordenskjöld reached an elevation of 5,000 feet at an estimated distance of 73 miles from the ice-margin; Peary (east of Disko) of 7,525 feet, at 100 miles; Nansen of 8,970 feet, at 112 miles; and Peary, in the far north, of 8-9,000 feet, at seemingly not more than 80-100 miles. Nansen assumes the curvature of the ice-cap to be the correspondent of an arc of a circle, and he attributes the outline to an evenly-timed or distributed movement which has been brought about by the weight (mass) of the ice-cap itself. While it may not be easy to disprove (just as it is impossible to prove) this proposition, it does not appear to me that there is any real evidence to support it. The inland contour is too flat to permit of any hopeful consideration of the cause of the curvature such as it is.

Seaward the fall of the ice-cap to its lower level is in places rapid. Beyond the Arctic Circle this lower-level—the line of perpetual snow—is held in a general way between 1,700 and 2,200 feet, and seemingly a latitudinal distance of ten or fifteen degrees does not materially disturb this position. We found the lower margin of the summer snows on Disko Island (Lat. 69°) at 1,800 feet, on the south side of McCormick Bay (77° 30′) at 2,200 feet, on the northeast angle of the same bay at 1,800 feet, and on the cliffs about Sonntag Bay (Lat. 78° 30′) at 2,000 feet. The exposed high ground beyond the 82d parallel sustains the evenness of physiographic conditions. At the northeast angle of McCormick Bay, where we made our search for the Peary party, we attained an elevation of 3,300 feet not further than eight miles from the ice-border, and 4,000 feet is reached within the next four or five miles. The ascent is here, therefore, approximately 180 feet to the mile, most of which is determined by the structural relief of the land.

The most pertinent inquiry regarding the ice-covering of Greenland is that relating to vertical development. It is commonly assumed that the thickness of the snow-cap cannot be less in places than 5-6,000 feet, and it is even thought that it might considerably exceed this figure. The reasons assigned for this hypothetical development are generally stated to be: 1, that the snow-surface has, in fact, an elevation of 8-10,000 feet, without any landmass (or true relief) rising through it; 2, snow falling through long ages must have accumulated to enormous depths; and 3, no valleys or deep depressions are anywhere to be met with in the interior (consequently they are filled in, and to an unknown depth). Naturally, this assumed thickness is purely conjectural, since there are no means for ascertaining its true measure; but the circumstance that no mountain ridges or natural orographic lines, indicative of a high relief of the interior, rise above the plateau-surface, would seem to lend color to this hypothesis. Assuming that such a vast deposit of snow does in fact exist, it becomes interesting to inquire into the conditions governing its accumulation and first formation.

We know nothing of the petrographic relief of the interior, but reasoning from analogy or from our knowledge of other large land-masses, there is reason to assume that Greenland conforms to the normal type-structure of having its greater eminences seaward, and the lower elevations central.* If this is true, then necessarily must the interior snows have a much greater development than the border snows, since the relief which supports them can occupy but a fraction of the full height of the 8,000–10,000 feet to which the surface of the plateau rises. But by what process of accumulation can the interior snows acquire a development so much greater than that of the marginal snows? One would naturally look to the sea-bord as the region of excessive or greatest precipitation, and, indeed, it is difficult to con-

---

* It can, however, be assumed that the whole interior is, like Mexico, one vast rock plateau of very nearly uniform elevation. While it is true that such a plateau of igneous rock may exist over a very large area north of the 69th parallel of latitude, it is equally true that over still larger areas it does not exist, and that its absence in no essential way modifies the relief of the land. Nansen, on the other hand, assumes that in all probability the configuration of the country is similar to that of Norway, or the Scandinavian peninsula—a mountainous country, with its greater elevations central. The comparison, however, can scarcely be said to apply. The Scandinavian peninsula, like Italy, is a narrow backboned strip, very different from the broad expanse of Greenland, which has many of its most elevated summits situated not far from the coast-border.

ceive how, in a region like Greenland, the interior can be more favored in this respect than the exterior. The depleting of the clouds by the cold coast-wall, and the nearness to it of the open evaporating basin, the sea, must almost of necessity cause a maximum discharge over the littoral tracts. The vast thickness of the inland snows and ice must be due to a cause other than that of simple or natural precipitation. Nordenskjöld assumed that the strong and rapid condensation of moisture by the peripheral tracts of Greenland could not but deprive the in-blowing winds of their vapor, and thereby create a region in the interior largely devoid of precipitation—a dry oasis. This was practically a restatement of the hypothesis enunciated by Alexander von Humboldt that, were the Alps only moderately higher than they now are they would pass into a region beyond the clouds, and consequently into a snowless plane. The rapid diminution of snow-fall on the Alps as we ascend from about 8,000 feet to 10,000 feet, which has been so significantly demonstrated by Dollfuss-Ausset and the Schlagintweits, certainly seems to sustain this position; yet it cannot be denied that the enormous thickness of snow and ice that is found at the greater elevations of some of the Alpine peaks—for example, 300 feet, on the Jungfrau at an elevation of 10,200 feet, of 200 feet on the

146 *The Greenland Ice-Cap and its Glaciers.*

Piz Bernina at 13,000 feet—is a condition which is not absolutely in harmony with that rapid diminution of precipitation which has been assumed. On the other hand, the fact that in many years a number of the more prominent Alpine peaks were deprived of their snowy covering, shows that at certain periods this covering is not very thick.* The condition of the interior of Grinnell Land, much of which—even the "hills" of respectable elevation, of from 1,500 to 3,000 feet—was found to be devoid of a snow-covering by the officers of the Greely Expedition, tends in a measure to confirm Nordenskjöld's hypothesis. On the other hand, it is well known that Nordenskjöld's own researches in the interior of Greenland, as well as the more recent observations of Nansen and Peary, failed to demonstrate the condition that had been assumed,

---

* It should be noted with reference to the observations made by Dollfuss-Ausset on the Théodule Pass, that probably many of the more elevated mountain-regions receive their heaviest snow-deposit during the summer months (instead of the winter), when the clouds, also more highly charged during this season of the year, ascend to the highest level before they are deprived of their vapor. The vastly diminished snow-fall that has been observed during the winter months in t e high northern regions and the low position of the cloud-line seem likewise to point to the conclusion that the snows of the loftier Greenland mountains, whose crests and summits rise to 8,000 or 10,000 feet, must be principally the result of summer, rather than of winter, accumulation. The mountain-peak snows must, however, be distinguished from the snows of the interior.

but on the contrary give a flat contradiction to it. The interior of the land harbors no oasis, but is one continuous ice (snow) sheet.

To explain the anomaly that despite the "drying" of the in-passing clouds by the marginal mountains the snow accumulation in the far interior should be much heavier than it is on the coastal fringe, it can be assumed that the interior receives the discharges from the clouds from *both* the water areas, the eastern and the western, and that what is lost in actual quantity from one side alone is more than made good by a remaining quantity received from the other side. There would then be an equality at least, if not a large excess. This may be true to an extent, and there is every reason to believe that there is a certain amount of overlapping of the cloud-banks coming in from the opposite sides. But granting the overlapping of the deposits from the two sides, it is still exceedingly questionable if it would explain the enormous excess of material as it appears to-day in the interior.

An easier solution seems to be given by assuming that the inland snows are largely a wind-drift accumulation—the winds driving in to the interior (and there in a measure focussing) necessarily promoting excess of accumulation. This is, in fact, the simple explanation of well-known and to an extent parallel phenomena which are presented in regions of formative sands and loose earth—the "sand-mountains" of many desert tracts, much of the loess of China, the high hills of coral islands, etc. It is well known that the entire height of the Bermuda Islands, 260 feet, is of wind-drift construction, and still greater elevations of the same kind have been recorded as occurring on other coral islands. It is true that these hills are in the nature of giant dunes rather than extended plateaus, but this is due mainly to the limited area over which wind-action has free play; were the areas of deposition much larger, they would almost positively conduce to plateau formation.

So far as this explanation, applied to the interior Greenland snows, is concerned, it may be contended that the prevalent winds of the region are largely from the contrary direction —*i. e.*, they blow from the ice-cap seaward, and ought consequently to produce diminution, rather than accumulation, in the interior. Mr. Peary found the out-blowing winds almost constant throughout his journey—indeed, they were in great measure his directors in a "compass" course. But the journey was accomplished in the summer months, and it could hardly be expected that the course of the winds would be otherwise in that season of the year. The warm sea-area is at that time the suction area, and to it would be directed the indraught of cold air from the ice-covered interior.

During probably full eight months of the year the reversed conditions would obtain, or rather, the direction of the winds would be reversed. The preponderance of ocean- over land-winds would give a large balance in favor of the in-blow with the result of causing the snows to drive inward, whether from the one side or the other, and there spread themselves in the form of a vast plateau-surface. This is, as I conceive it, the most likely explanation of Greenland's heavy ice-cap, a parallel to which is presented by the sand and ash plateau of the Mexican Republic.

What thickness of the interior ice and snows is really needed for the formation of the large glaciers which almost everywhere push out into the direction of the sea, is a question that cannot, perhaps, be answered at the present moment. It is all but certain, however, that the enormous thickness which has generally been insisted upon is not a necessary condition for the existence of the large ice-sheets. Indeed, apart from latitudinal extent, it is doubtful if the thickness of the ice in any of the Greenland glaciers exceeds, or even equals, that of the largest ice-sheets of Switzerland. The Humboldt Glacier, which discharges beyond the 79th parallel into the Kane Basin (Smith Sound), and with a frontage stated to be from 50 to 70 miles, is generally assumed to be the largest of all known glaciers, yet the height of its sea-wall has been determined to be not much over 300 feet, or but little in excess of that of the Muir Glacier, and scarcely a fourth of the thickness of the ice which has been assumed for some of the Alpine glaciers.* None of the glaciers which I had personally the opportunity to examine had any really very great thickness of ice—*i. e.*, a thickness much exceeding 200—300 feet. Usually the seawall did not measure more than one-half this thickness. It is true that our examination did not extend more than a few miles up

* I strongly suspect that the dimensions of this glacier have been largely over-stated, as the dimensions of almost everything Arctic—capes, bays, islands, ice-bergs, dangers—are over-stated by the earlier explorers. The dimensions are so greatly in excess of those of any other Greenland glacier that a re-measurement should be made before they are accepted as a finality. The charts of the northwest coast are so erroneous in their details that little dependence can be placed on them: bays and sounds are given with double the extent that really belongs to them—*e. g.*, Whale and Murchison Sounds, Wolstenholme Sound, etc.,—and so their entering glaciers are made to appear very much larger than they really are. Many of the so-called *Mers de Glace* which appear on nearly all charts, leading one to assume the existence of vast interior or summit glaciers, are in reality only the ice cap or feeding-basins, the correspondent of the firn or névé. Again, where many maps indicate a solid land-frontage, without a glacial break, the actual face of the country shows it be almost riddled by glacial valleys. Thus, between Cape York and the Petowik Glacier no less than eight glaciers, and of goodly size, break through the granite cliffs, and the same configuration of glacier following rapidly upon glacier presents itself almost everywhere beyond the Petowik Glacier.

stream, and it is possible, and even very probable, that a considerably greater thickness of ice would be found (in some of the glaciers, at least) in positions nearer to their feeding basins. The difficulty of approach to the Melville Bay glaciers, which are probably the largest after the Humboldt Glacier, prevented any careful observations being made on these, but seemingly even their terminal walls were of no extraordinary height. Yet it is true that the ice-bergs which course about Melville Bay are many of them fully 200—250 feet above water, and, doubtless, they descend in cases at least 600—800, or even a thousand, feet beneath the water-line. In what manner the height of an ice-berg is related to the thickness of glacial ice, is a question that still remains to be determined. The height of a berg may very readily be a length in section of a glacier, especially in the case of discharges from the solid-bodied or non-crevassed ice-streams.

The insignificant development of the ice-cap, in its relation to the large glacial streams which radiate off from it, is so striking a peculiarity in some parts of Greenland as to have suggested the suspicion that many of the existing glaciers are relicts of the Great Ice Age. Bessels, the accomplished scientist of the Polaris Expedition, actually assumed this explanation for the by no means insignificant glaciers of Herbert and Northumberland Islands, lying somewhat above the 77th parallel of latitude, which descend from an ice-cap of from 1,800 to 2,500 feet elevation; it did not seem possible that the comparatively feeble accumulation of snow which is found at this altitude could originate ice-streams of the dimensions which are there found. I am convinced that there is no real basis for this interpretation. The snow-covering of these islands belongs to themselves, and feeble though it be, it is quite competent to explain the associated phenomena. Many of the "hanging" glaciers of Herbert Island, which descend over slopes of some 30—35 degrees, are so attenuated in their upper parts as to be almost extinguished before reaching the summer ice-cap; yet basally they rapidly increase in dimensions, so that before they finally disappear they measure not less than 40—50 feet in thickness. The slowly accumulating snows descend over the first-formed layers, whether by sliding or otherwise, and help to build up the base while they thin out the top. These hanging glaciers have all the structural features of the normal glacier type. I am convinced that some of the minor glaciers have, indeed, been formed without the assistance of any ice-cap or of the accumulated snows of a névé basin. Such glacial streams might, perhaps, be advantageously termed ravine or *couloir* glaciers; they seem not to be numerous.

Geologists frequently distinguish between the glaciers of Greenland and those of other countries, and insist that the phenomena of the Great Ice Age must be studied in the light of the evidence afforded by the ice-sheets of Greenland (and of other Arctic regions) rather than by that of the glaciers of south-central Europe, of the Caucasus, or of the Himalayas, I have not been able to determine precisely in what the distinction lies. The glaciers (and their workings) of Greenland seem to me to be the exact counterpart of those of the Alps; the differences which they present are individual and not specific, just as they are in all regions of glaciation. They are large or small, are steeply-inclined or nearly horizontal, rapidly-moving or almost stationary, just as special conditions of formation and moulding present themselves. They may, again, be largely fissured or solid, provided with moraines or destitute of them, and clean or dirty. Many of them discharge into the sea, others stop before reaching the water-line. All are now, or have been in a recent period, undergoing contraction (recession).

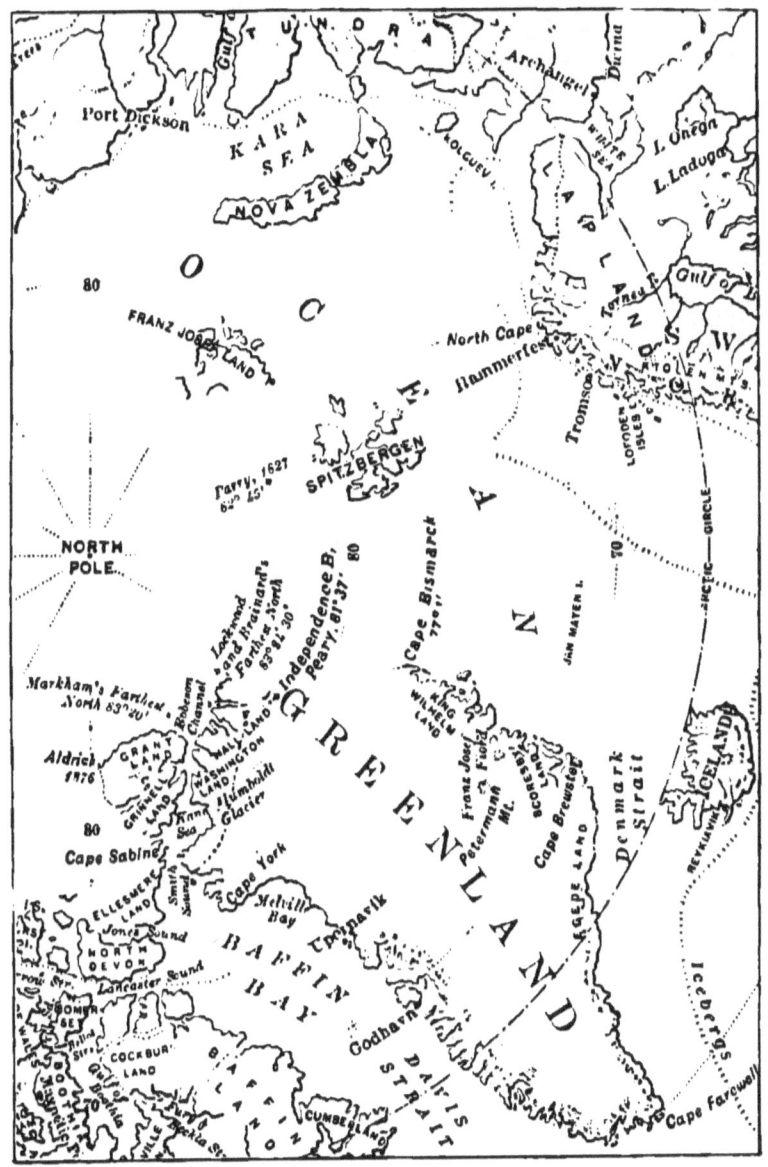

INDEX OF REFERENCES.

Name.	Page.
Aldrich,	11
Andersson,	95
Astrup,	81, 107, 125, 133
Back,	54, 56
Baflin,	33, 34
Barentz,	28
Barrington,	25
Barrow,	11, 30, 47, 53
Bateson,	34, 35
Belcher,	54
Bessels,	84, 103
Beyer,	96
Brainard (See Lockwood and Brainard.)	
Broberg,	95, 135
Bruyne,	59
Bryant,	84, 99, 118, 123, 135
Buchan,	37, 38, 58
Buddington,	54
Bylot (See Baflin.)	
Clarke,	34, 35
Collinson,	54, 56
Cook,	81, 103, 107, 129, 133
Cordiner,	16
Crawford,	54
Dollfuss-Ausset,	145
"Dominus Vobiscum,"	28
Dunphy,	98, 107
Ekroll,	8, 9
Entrikin,	84, 99, 124, 135
Findlay,	56
Fitzroy,	55
Fotherby,	33
Franklin,	13, 37, 38, 53, 58
Frederick,	90, 92

Frobisher, ... 17
Garlington, ... 80, 84
Gibson, ... 81, 107, 133
Gray, ... 59, 70
Greely, ... 14, 67, 80, 146
Hall, ... 14, 49, 50, 104
Hayes, ... 49, 103, 104
Henson, ... 81, 107, 119
Hite, ... 84, 135
Hudson, ... 30, 31, 32
Humboldt, ... 145
Inglefield, ... 55
Jackson 73-75
"Jeannette," 7, 18
Kane, ... 49, 84, 104
Koldewey, ... 31, 54, 59, 70
Lamont, ... 55
Lockwood and Brainard, . . 10, 13, 20, 21, 22, 46, 63, 72
Mac Callam, ... 25
Markham, A. H., ... 14, 19, 22, 67, 72, 73
Markham, C. R., ... 54, 56, 60, 73
Maury, ... 54
McClintock, ... 54, 56, 63, 64, 85
McClure, ... 13
McKinley, ... 129
Meehan, ... 84, 135
Melville, ... 67
Mills, ... 84, 199
Murchison, ... 55
Murphy, ... 129
Nansen, ... 7, 141, 142
Nares, ... 49, 58, 61
Nordenskjöld, . 13, 16, 56, 57, 58, 65, 66, 91, 141, 145, 146
Ommaney 54, 56
Osborn, ... 54, 56, 60
Palander, ... 65
Parr, ... 19

Parry, 8, 13, 14, 22, 23, 24, 27, 37, 39-51, 58, 62, 63, 65, 67
Payer, ... 21, 22, 31, 59, 60, 65
Peary, R. E., 10, 21, 22, 46, 70, 71, 81-84, 107, 112, . . .
. . . . 125, 129, 135, 136, 139, 148
Peary, J. D., ... 81, 107, 111, 119, 126
Petermann, ... 51, 54, 58, 59, 60, 65, 70
Peterson, ... 16
Phipps, ... 33, 34, 35, 36
Pike, ... 80, 84, 85, 110
Poole, ... 32, 33
"Proteus," 80, 84, 114
Rae, ... 56, 64
Rawlinson, ... 55
Richards, ... 52, 54, 56, 64
Robinson, ... 26
Ross, J., ... 37, 103
Ross, J. C., ... 58
Roule, ... 60
Ryder, ... 70
Ryp, ... 28, 29
Sabine, ... 54
Schlagintweit, ... 145
Scoresby, ... 34, 35, 38, 53
Smith, ... 60
Stephens, ... 26
Stokes, ... 84, 135
"Tegethoff," 18, 59
Verhoeff, ... 81, 107, 128, 129, 138, 139
Vorse, ... 84, 135
Weyprecht, ... 17, 31, 58, 59
Wiggins, J., ... 16
Wiggins, R., ... 16
Young, ... 55, 74

CORRECTION.

Pages 7, 18. For Jeanette read Jeannette.

Captain Richard Pike--A Retrospect.

While these pages are going through the press intelligence is received of the death, on the 5th of May, of Captain Richard Pike. Through this death the guild of the northern "ice-masters" is deprived of one of its most noted and conspicuous members. From early manhood through to his last day Captain Pike was a man of the sea. Although not strictly fond of the "ocean blue," he was restless on land, and seemed always impatient to be active in his calling. The vessels under his command sailed the waters from Newfoundland to La Plata, and from the Mediterranean to the frozen north. He was thus a man of all climates, but his home, as a sealer and whaler, was preeminently the abode of snow and ice. His familiarity with the northern waters was such as to always insure respect and consideration for his opinions and judgment, a confidence that was not shaken through the misfortunes which broke into his career, and which a long period of active service is almost certain to compass. The recollection of one of these seems never to have completely left the mind of the good and genial tar, no more than the reality ever left the body in which it had sown the seeds of dissociation. The crushing of the Proteus, the vessel of the Second Greely Relief Expedition, by the ice-floes of Smith Sound in July, 1883, had graven a deep reminiscent furrow—one too deep even for the mellowing influences of time to efface; nor did exoneration by two Naval Boards of Inquiry succeed in restoring to him that spirit which was shattered in the thought that the fate of an Arctic expedition hung in the balance of his misfortunates. But Captain Pike remained true and good to his men, and to the rugged Newfoundlanders he was endearingly known as "good old Captain Pike."

The following pages briefly recite the events of a few days surrounding the catastrophe to the Proteus—the approach to Cape Sabine, the crushing of the vessel in the ice, and the preparations for the retreat. They are part of the official "log" of the vessel which, was confided to the author by Captain Pike, with the request that it might some day be given to the public as a supplement to an Arctic journal. No more fitting place appears than the narrative of the expedition of the Kite, the ship which he so successfully conducted to her point of destination on two successive voyages.

Thursday, July 19th:—A. M. begins moderate, breeze from S. E., with rain and thick fog—5 A. M., fog clearing up, no water to be seen to the northward—hauled ship back to S. E., towards some lakes of water, thinking to get around the ice in that direction—ship steaming full speed—7, clear, saw the land—noon, fine, no water to be seen along the coast. Latitude by meridian altitude 75° 30′ N.—3 P. M., stopped the ship by the edge of ice off some islands in Melville Bay in Lat. 75° 42′ N., Long. 61° 50′ W.—7 P. M. saw no chance to get to Cape York in shore, turned ship and went full speed to the southward to try to get around to the westward—Midnight calm and clear sky, a little young ice making in the lakes of water; ship making good way through loose ice to the westward.

Friday, July 20th:—A. M. begins calm and clear—ship steaming full speed around large sheets of ice in a westerly direction—2 A. M. shot a Polar bear after chasing him some time —3 A. M., ice close, but ship making good way to the westward, two men belonging to the expedition busy skinning the bear—8 A. M., ice in very heavy sheets to the westward, turned ship to the southward towards some loose ice. —Noon, got in a southern water which trimmed around to N. W., going full speed, Lat. obs. 75° 17′ N.—2 P. M. turned after a Polar bear, shot him after a long chase, hoisted him on board and went full speed on her course—7 P. M., Cape York S. E. about 15 miles, light breeze from S. S. E., with light misty rain—8.30 P. M. shot a dog hood-seal on the ice, hoisted it on board and skinned it—Midnight calm and clear, but very thick to the southward; going full speed along the coast, very little ice close in, great deal to the westward.

Saturday, July 21st:—A. M. begins calm and fine—1.30 A. M. passed Conical Rock off Cape Dudley Digges about a mile off, passing through loose ice towards Carey Islands, as we were going to visit the eastern one of the group —7 A. M. off Cape Athol, thick, black fog to the northward, close down to the water, but clear along the coast to the eastward.—8 A. M. thick fog all around, ice in very large sheets but not heavy—10 A. M. fog cleared off, no chance of getting north on the course steering, as it was all one sheet between the ship and the island, turned ship to S. W. in some veins of water to get around to the westward, strong wind from W. S. W.—Noon, lakes of water trimming north, followed around the sheets S. E., Carey Islands bore N. E.—3 P. M. hove to, off the island in open water—bergs very thick, landed on the island to examine stores landed there by the English Expedition in 1875—6 P. M. boat returned, reported the boat left there in good condition, but a great deal of the stores spoiled—6.30 P. M. turned ship northwards, going full speed for Pandora Harbor—11 P. M. passed Hackluyt Island—Midnight, light breeze from S. W., and fine but hazy on the land.

Sunday, July 22nd:—A. M. begins calm and fine, going full speed to the northward in open water—5 A. M. close by Cape Alexander, stopped the engines a short time for some little repairs—when finished, started again for Pandora Harbor—7 A. M. steamed in the Harbor, Expedition party went on shore to see for the record left by the S. S. Neptune in 1882, but failed to find any—8 A. M. boat returned, steamed on for Littleton Island, stopped there a little while, but did not land, started full speed to the northward—Noon, met ice to the north, steamed to the edge—not a break to be seen to the northward; steamed to the edge, still all solid north—turned ship towards Cape Sabine—3.30 P. M. Lieut. Garlington went to the Cape to look after a depot of provisions left last year by the S. S. Neptune—7 P. M. came on board and reported open water to the northward, had some conversation with him as to going out in the ship—7.30 hove up anchor and proceeded full speed toward the Cape, saw a little water across toward Cape Albert and by hard steaming and butting sheets to the north gained that point.*—Midnight, water closing up still . . . . north toward Cape Albert—calm and fine.

Monday, July 23rd begins calm and fine—

---

*The last three words are erased in the log.

2 A. M. butting for open water off Cape Albert, but could not succeed; turned ship in an easterly direction to try to get around; got within about 100 yards of the water when the ship got nipped and could not move off Cape Albert—5 A. M. ship loose, steamed north about 5 miles when we were stopped; no water to be seen north; put back to the open water south towards Cape Sabine—7 A. M. lay up in a small lake of water—8, started south again, at times very difficult to get the ship through as the ice was very close and heavy—3 P. M. close by the main water in the Sound when two heavy sheets caught the ship, rafting all around her, carrying starboard, main rail, bulwark, stanchions and everything before it; ordered boats and provisions to be put on the ice which was still rafting—4.30 ice stove in the starboard side abreast of boiler—6.30 P. M. filled to the decks with water—7.30 ice opened and the ship sank; Cape Sabine S. S. E. about 10 miles; held a consultation with Lieutenant Garlington and Colwell regarding getting provisions, etc., landed.—Midnight started one boat, being successful in landing one load a little to the west of Cape Sabine.

Tuesday, July 24th.—A. M. begins light breeze from the westward and fine—ice drifting out around Cape Sabine very fast; all hands employed with five boats landing provisions off the ice—8.30 A. M. all hands left the floe and landed close to Payer Harbor where most of the provisions and clothing were landed—2 P. M. boiled some tea and had dinner; the rest of the afternoon employed fixing up stores and collecting them together; wind south with thick fog—a little ice passing down the sound; midnight, strong breeze with thick fog.

www.ingramcontent.com/pod-product-compliance
Lightning Source LLC
Chambersburg PA
CBHW032134160426
43197CB00008B/633